Green Energy and Technology

Panagiotis Grammelis
Editor

Solid Biofuels for Energy

A Lower Greenhouse Gas Alternative

Springer

Panagiotis Grammelis, PhD
Centre for Research &
Technology Hellas (CERTH)
Institute for Solid Fuels
Technology & Applications (ISFTA)
357–359 Mesogeion Ave.
Halandri
152 31 Athens
Greece
grammelis@certh.gr

ISSN 1865-3529   e-ISSN 1865-3537
ISBN 978-1-84996-392-3   e-ISBN 978-1-84996-393-0
DOI 10.1007/978-1-84996-393-0
Springer London Dordrecht Heidelberg New York

British Library Cataloguing in Publication Data
A catalogue record for this book is available from the British Library

Library of Congress Control Number: 2010936094

Cover design: eStudioCalamar, Girona/Berlin

Printed on acid-free paper

Springer is part of Springer Science+Business Media (www.springer.com)

Preface

Fossil fuels are widely used for electricity generation and heating, emitting green-house gases and other toxic pollutants, which should be minimized according to the most recent environmental legislation. The utilization of solid fuels of biogenic origin could contribute to some extent towards this aim. Although a lot of information is available for liquid biofuels, only few data can be found for the research trends in the solid biofuels sector. The scope of this edition is to present the current status of the engineering disciplines in the specific area, providing an extensive overview of the energy exploitation options of solid biomass. In this sense, all thematic priorities related to the solid bioenergy potential and standardization, energy technologies – commercialized and emerging ones – and quality of solid residues are presented. Special attention has been given to biomass co-firing with coal, since it has the highest potential for commercial application in large-scale units, whilst combustion and gasification are more promising for units of small to medium scale. Key aspects for the energy exploitation of solid biofuels are considered in this book, providing valuable information for the reader who is familiar with the biomass sector. Even for an amateur, basic knowledge is provided, since all potential methods for solid biomass exploitation are described.

More specifically, in Chapter 1 efforts are made to address key questions arising for biomass availability and supply. The bioenergy supply potential has already been assessed at global level by the IPCC, US EPA, World Energy Council, Shell, IASA, and the Stockholm Environment Institute. However, estimates of the biomass share in the future global energy supply range widely from below 100 EJ/year to above 400 EJ/year in 2050. Thus, it was attempted to provide a detailed analysis on solid agricultural biomass feedstocks in EU27 and to summarize relevant data that influence the availability and future supply of these feedstocks for energy and fuel production.

The European Standards for the specifications and classes of solid biofuels are presented in Chapter 2. Among the 27 technical specifications for solid biofuels, the classification and specification (CEN/TS 14961) and quality assurance (CEN/TS 15234) are the most important and are upgraded to full European standards (EN). Part 1 – General requirements of EN 14961-1, which is the objective of this chapter, includes all solid biofuels and is targeted for all user groups. The classification of solid biofuels is based on their origin and source and biofuels are divided to four sub-categories: (1) woody biomass, (2) herbaceous biomass, (3) fruit biomass, and (4) blends and mixtures.

Chapter 3 provides an overview of all technical issues for biomass–coal co-firing in boilers designed exclusively for coal – mainly pulverized – combustion. Biomass–coal co-combustion represents a near-term, low-risk, low-cost, sustainable, renewable energy option that promises effective reduction in CO_2, SO_x and often NO_x emissions, as well as several societal benefits. Technical issues associated with co-firing include fuel supply, handling and storage challenges, potential increases in corrosion, decreases in overall efficiency, ash deposition issues, pollutant emissions, carbon burnout, impacts on ash marketing, impacts on SCR performance, and overall economics.

A step ahead on co-firing development is covered in Chapter 4, in which the co-utilization of Solid Recovered Fuels (SRF) with coal is extensively reviewed. SRF are solid fuels prepared from high calorific fractions of non-hazardous waste materials intended to be fired in existing coal power plants and industrial furnaces. The use of waste as energy source is an integral part of waste management. As such, within the framework of the European Community's policy-objectives related to renewable energy, an approach to the effective use of wastes as energy sources is outlined in documents like the European Waste Strategy. The scope of this chapter is to cover the SRF characterization using the pre-nominative technical specifications of CEN/TC 343 and the status of the standardization activities. Additionally, some of the experiences gained from co-firing of SRF and biomass in large scale demonstration plants is summarized. These include handling and pre-treatment of the SRF, milling corrosion, emissions behavior, and the quality of solid residues.

The subject of Chapter 5 deals with biomass combustion characteristics. Unlike pulverized coal, biomass particles are neither small enough to neglect internal temperature gradients nor equant enough to model as spheres. Experimental and theoretical investigations indicate particle shape and size influence biomass particle dynamics, including essentially all aspects of combustion such as drying, heating, and reaction. This chapter theoretically and experimentally illustrates how these effects impact particle conversion.

Fluidized bed combustion (FBC) technology developed in the 1970s was soon expanded for biomass and other low-grade fuels, as presented in Chapter 6. The benefit of the FBC is the large amount of bed material compared with the mass of the fuel and, thus, the large heat capacity of the bed material that stabilizes the energy output caused by variations in fuel properties. Moreover, by selecting reagents as bed material and controlling the bed temperature, the emissions of pollutants can be controlled. In the last two decades, rapid progress has been achieved in the application of FBC technology to power plants up to intermediate capacities, caused by the increasing demands for fuel flexibility, stringent emission control requirements, stable plant operation, and availability. The main objective of this work is to review critically the technical requirements of biomass and/or waste combustion in FBCs, the operational problems, the needs for emissions control, and the ash handling issues.

Another thermochemical conversion technology for biomass is gasification, which is examined in Chapter 7. Gasification is a mature technology for energy production that permits an easier separation of CO_2 for its storage. Wastes gasification reduces the dependence on fossil fuels and co-gasification with coal could provide the benefit of security in fuel supply, as the availability of wastes and solid biofuels varies from region to region and demonstrates seasonal changes. Gasification experimental conditions and technologies and syngas cleaning methods are key issues for the production of a clean gas that could find a wide range of applications. This chapter concentrates on syngas end-uses, focusing on new ones, like gas turbines or engines in IGCC, synthesis of methanol, ethanol, and dimethyl ether, Fischer–Tropsch synthesis, and hydrogen production.

Integrated schemes of micro-CHP and biofuels are very promising for decentralized applications. Renewable micro-CHP systems are a combination of micro-CHP technology and renewable energy technology, such as biomass gasification systems or solar concentrators. The integration of renewable energy sources with micro-CHP allows for the development of sustainable energy systems with the potential for high market penetration, a cost-effective and reliable heat and electricity supply, and a highly beneficial environmental and economical impact on a pan-European scale. Chapter 8 discusses the state of the art technological options in the field of renewable micro-CHP with biofuels with regards to technology, cost, and environmental impacts, and presents a market survey concerning the possibility of future penetration of the technology in Europe. The results provide a coherent overview of the basic technological options for renewable micro-CHP with biofuels and provide an insight on the market trends within Europe and projected future market scenarios, taking into account cost estimations for various micro-CHP technologies, feedstocks, and electricity and fuel prices in Europe.

Chapter 9 provides an overview of the main ash formation and deposition mechanisms for various relevant biomass fuels, also in blends with selected coals, in pulverized-fuel (PF) boilers. The chapter is divided into three sections. In the first, a general outline of the ash formation mechanisms is described. The second section includes a review of experimental and analytical techniques for the lab-scale characterization of fuels, emphasizing the ash-forming elements contents and fate during combustion. In the third section, key ash-formation phenomena are discussed for various pure biomass fuels and selected typical coals, which is based on exemplified results, generated with the techniques discussed in the foregoing section.

The book ends with an overview of the different forms of ash utilization that exist or are being developed for biomass ashes, as presented in Chapter 10. The first section reviews options for ashes from biomass co-firing with coal, both established forms of utilization in cement and concrete, and alternative options, *e.g.*, manufacture of lightweight aggregates. The second section discusses utilization options for residues from "pure" biomass combustion. The large variation in biomass fuels and installation types makes this a complex issue. Besides recycling of clean wood ash to forests, these are all emerging forms of utilization. The third section deals with the specific issues related to the utilization of carbon-rich ashes from biomass gasification and pyrolysis.

Acknowledgments I would like to thank all authors who assisted in accomplishing the objectives of this edition and providing a detailed and in-depth analysis of the solid biofuels sector. Special thanks to my secretary, Mrs Angeliki Diafa, for taking care of all editing details up to finalization of the book.

Panagiotis Grammelis

Contents

Chapter 1
Supply of Solid Biofuels: Potential Feedstocks, Cost and Sustainability Issues in EU27

Calliope Panoutsou

Abstract In 2006, the total biomass contribution to primary energy consumption in the European Union was 86.6 million tons oil equivalent (Mtoe). The main share of 66.4 Mtoe was provided by solid biomass, with the remainder provided by biogas, transport biofuels and renewable solid municipal waste [1]. The bioenergy supply potential has recently been assessed at global level by (among others) the IPCC, US EPA, World Energy Council, Shell, IASA and the Stockholm Environment Institute [2, 3]. Estimates of the share of biomass in the future global energy supply range from below 100 EJ/year to above 400 EJ/year in 2050, compared to a global primary energy consumption of 420 EJ for the year 2001 [4]. One of the major reasons for the large ranges observed is that studies differed widely in their estimates of land availability and energy crop yields, and, to a lesser extent, the availability of wood and residue resources. Studies at the European level also deliver widely ranging results. Conservative results on the total biomass potential come from the EEA study [8]: how much bioenergy can Europe produce without harming the environment? It estimates a total bioenergy potential from agriculture, forestry and waste of almost 300 Mtoe in 2030. Of this, 142 MTOE will come from agriculture only which is obtained from 19 million hectares of agricultural land. This is equivalent to 12% of the utilised agricultural area in 2030. The purpose of this chapter is to provide a perspective on solid agricultural biomass feedstocks in EU27 and to summarise relevant data that influence the availability and future supply of these feedstocks for energy and fuel production. To achieve this, the chapter is structured in sections that aim to provide a series of concise answers to key questions arising regarding biomass availability and supply:

C. Panoutsou (✉)
Imperial College London,
Centre for Energy Policy and Technology (ICEPT), Mechanical Engineering,
Exhibition Road, London SW7 2AZ, UK
Tel: +44 207 594 6781 Fax: +44 207 594 9334
e-mail: c.panoutsou@imperial.ac.uk

1

- Which types of biomass feedstocks can be produced within the available land resources of EU27 and how much of them can be estimated as available.
- What are the key cost factors and the costs ranges for residual feedstocks and energy crops.
- What are the main concerns affecting their sustainable exploitation.
- What are the main future challenges and how they can be overcome.

1.1 Biomass Feedstocks

Biomass feedstocks can be categorized as agricultural, forestry and waste-based. Within each category there are primary sources, or material directly produced from current operations, secondary sources, derived from agro- and wood industries, and waste sources, from construction, demolition and municipal solid waste.

Agricultural Feedstocks
Primary

Field crop residues: field crops produce two types of field residues, *i.e.* dry and fresh or green residues. Green field crop residues, such as sugarbeets, potatoes, onions, *etc.* are left in the field in fresh, succulent condition. These residues have high moisture content, usually more than 70%, are usually rotting in the field and rarely are any of them used for animal feeding [5].

The main bulk of field crop residues are left in the field in a semi-dry or dry condition. Dry field residues are derived from field crops, cultivated in various EU regions, and they may come from small grain cereals (wheat, barley, oats, rye and rice), maize, oil crops (sunflower, rapeseed, *etc.*), cotton, tobacco, *etc.* These residues are incorporated into the soil, burned in the field or collected and used for energy and various other purposes.

Availability of crop residues depends primarily on the choices of crop and the requirements for food and fodder production. The availability of field-based residues depends on residue to product ratios as well as crop production and management systems.

Most of the studies considered by this review assumed that about 25% of the total available agricultural residues can be recovered [6–9]. Hall [10] estimates the potential of agricultural residues to be in the range 14 EJ/year and 25 EJ/year. The potential contribution of crop residues is assessed by Lazarus [9] to be 5 EJ/year. Fischer and Schrattenholzer [3] have assessed the crop residue potential for five crop groups: wheat, rice, other grains, protein feed and other food crops. The contribution of crop residues is 27 EJ/year in their high potential assessment and 18 EJ/year in their low potential assessment – similar values to Hall [10]. Hence, the range of primary agricultural residues included in this study varies between 5 EJ/year and 27 EJ/year.

Secondary

Agro-industrial residues: secondary or process-based residues are residues obtained during food processing, like bagasse and rice husk. This has to be derived from the production of crops that produce valuable secondary residues and from the residue fraction available after processing these crops. Concerning bagasse, it is assumed that all of the produced quantities can be recovered and used for energy applications [6, 8–10]. Based on these assumptions, the total potential of secondary residues is assessed at 5 EJ/year.

Animal wastes: two main sources of animal residues are manures and slaughter residues – the latter is not included in this review.

Manure availability depends on the number of animals and the use of manure for fertilizer. Wirsenius assessed the total amount of manure produced to be 46 EJ/year [34]. Several studies have assumed that 12.5% [10] to 25% [6–8, 11] of the total available manure can be recovered for energy production. With Wirsenius' [12] figures, the net available amount would hence be 6–12 EJ/year. However, these figures are for the period 1992–1995, while other studies included the growth of animal husbandry. Other estimates for manure production give figures of 25 EJ/year [4] and 13 EJ/year [13] available for energy use.

Hence, the availability of energy from animal manure included in this study ranges from 12 EJ/year to 25 EJ/year, depending on rate of growth of animal husbandry and rate of recovery *vs* fertilizer use.

Forest Feedstocks
Primary

The average area of forest and wooded land per EU Member State varies regionally (Figure 1.1). The area varies between 27.6 million ha in Sweden and 117 ha in Cyprus.

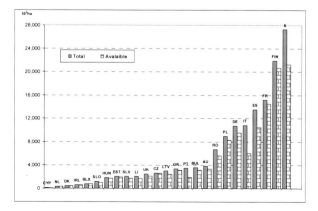

Figure 1.1   Total and available forest land in EU27

Table 1.1 Forest biomass per EU region

EU Region	Woody biomass available			Wood management		
	Growing stock	Stump and roots	Fellings (F_j)	Removal (R_j)	Forest residues (F_j–R_j)	Available residues (R_{fj})
[–]	[$m^3 \times 10^3$]	[m.t.$\times 10^3$]	[$m^3 \times 10^3$]	[$m^3 \times 10^3$]	[$m^3 \times 10^3$/year]	[m.t.$\times 10^3$]
Central	10,759,064	5,602,685	231,296	183,567	47,729	19,887
South	1,943,301	1,044,755	43,248	34,629	8,619	3,591
North	6,295,875	3,346,720	147,765	124,067	23,698	9,874

Potential contribution of wood to energy supply varies accordingly. There are also large regional differences in accessibility to forests [14].

According to FAO [14], changes in forest area are mainly caused by afforestation of former agricultural lands aiming to increase long-term timber supply, to increase the level of non-wood goods and services and to provide alternatives for agricultural use of land. In general, afforestation activities have slowed down considerably since 1980. It is becoming harder to find land suitable for afforestation, and the costs of additional afforestation are increasing.

Forest residues are the residues left during onsite wood management activities. In order to have a better appreciation of the regional distribution of forest potentials we categorised EU27 countries in three major sub-groups, namely central (AU, FR, DE, NL, BUL, CZ, HUN, POL, RO, SK), north (S, FIN, EST, DK, UK, IRL, LV, LT) and south (GR, IT, PT, ES, SL, CYP) regions and presented data analysis accordingly (Table 1.1).

Secondary

Wood waste: a comprehensive review for "mill-site-generated wood waste" is provided by Parikka [15]. Approximately 45–55% of the log input to a sawmill or plywood plant becomes waste [13, 16, 17]. The actual quantity of residues differs from plant to plant and depends on several factors, such as wood properties, type of operation and maintenance of the plant [17]. Average figures for each type of industry are presented in Table 1.2.

Table 1.2 Residues generated (%) in selected forest products industries[a]. Source: adapted from Parikka [15]

	Sawmilling[b] (%)	Plywood manufacture (%)	Particleboard manufacture (%)	Integrated operations (%)
Finished product (range)	45–55	40–50	85–90	65–70
Finished product (average)	50	47	90	68
Residues/fuel	43	45	5	24
Losses	7	8	5	8
Total	100	100	100	100

[a] Excluding bark
[b] Air-dried

Energy crops: "energy crops" may be defined as crops specifically cultivated to produce biomass to serve as energy vectors to release energy either by direct combustion or by conversion to other vectors such as biogas or liquid biofuels [18].

Wastes: availability of organic waste for energy use depends strongly on variables like economic development, consumption pattern and the fraction of biomass material in total waste production. Several studies have considered the theoretical availability of organic waste for energy purposes (Table 1.3).

Table 1.3 Agronomic aspects of selected energy crops under study in EU

Species	EU regions	Sowing/establishment	Harvest	Yield (t/ha)	Remarks
Oil crops					
Rapeseed	Central, Eastern, Mediterranean	March–May	June–July	3–5 (grain)	Both annual (spring-sown) and biennial (winter-sown) types of Brassica napus ssp. oleifera are cultivated. Winter crops can be harvested from late July, spring ones usually ripening during September
Brassica carinata	Mediterranean	March–May	June–July	2.5–3 (grain)	Originated from Ethiopia, it is a late growing plant, which is mainly grown in warm tropical regions. It is vulnerable to the cool regions, thus its cultivation is not recommended to those areas with heavy winters
Sunflower	Mediterranean	March–May	June–July	2.5–4	Annual plant with a strong taproot, from which develop deeply-penetrating lateral roots. Modern crop cultivars may be less than 1 m tall (dwarf types) or 1.5 m (semi-dwarf) at maturity
Sugar crops					
Sweet sorghum	Mediterranean	March–May	Sept–Nov	16–35 (fresh stems) 14–20	C4 annual grass with a well-developed root system and robust aerial parts, which are usually supported by prop roots. Growth characteristics are very variable, depending upon the type; some varieties may exceed 4 m in height, while others may attain only 50 cm
Sugarbeet	Central, Eastern, Mediterranean				Annual crop requires good-quality land. High productivity and also higher emission levels of agrichemicals. Deployment in the UK, Germany and other member states for bioethanol production

Table 1.3 (continued)

Species	EU regions	Sowing/ establishment	Harvest	Yield (t/ha)	Remarks
Starch crops					
Wheat	All	March–May	June–July	2.5–9 (grain)	Wheat and barley are annual grasses 60–120cm tall. Varieties have been traditionally bred for starch and straw has been used for feeding and bedding purposes. Recently both crops are used as feedstocks for bioethanol production in Europe and worldwide
Barley	All	March–May	June–July	2–7 (grain)	
			June–July	10–15 (grain)	
Corn	All	March–May			Corn is recently used as a bioethanol feedstock with specific varieties being bred for this purpose
Lignocellulosic					
Fiber sorghum		March–May	Sept–Nov	16–27	A hybrid deriving from grain and broomcorn sorghums. Annual plant, growing to 3.5–4 m tall, with high water use efficiency. It can be grown successfully on a wide range of soils except water logged and acidic
Cardoon	Mediterranean	Feb–March or Sept–Oct	July–Sept	10–22	Low input, high biomass yielding crop, well adapted to the semi-arid Mediterranean climatic conditions. due to its winter growth and to its robust rooting system, it offers protection against soil erosion in sloping and marginal lands
Miscanthus	Central, Eastern, Mediterranean	March–June	Feb–April	12–24	Perennial C4-crop that is harvested each year. So far, only limited commercial experience in Europe. Breeding potential hardly explored
Giant reed	Mediterranean	March–May	Feb–April	12–24	C3 perennial crop, native to the Mediterranean region. Tolerant to various soil types with high productivity under irrigation. It abundant root system provides tolerance to drought conditions, efficient water uptake and protection to soil erosion

Table 1.3 (continued)

Species	EU regions	Sowing/ establish- ment	Harvest	Yield (t/ha)	Remarks
Lignocellulosic					
Switchgrass	Central, Eastern, Mediter- ranean	April– May	Feb– April	10–20	Perennial C4-crop that is harvested each year. It is a cool-season grass and does best on moderately deep to deep, somewhat dry to poorly drai- ned, sandy to clay loam soils. It does poorly on heavy soils
Willow	Central, Eastern, Mediter- ranean	March– April	Nov– Dec	8–20	Perennial crop with typical rotation of some 3–4 years. Suited for colder and wetter climates. Commercial experience gained in Sweden and to a lesser extent in the UK and some other countries
Poplar	Central, Eastern, Mediter- ranean	March– April	Nov– Dec	8–18	Perennial C3-crop, currently espe- cially planted for pulpwood produc- tion in various countries. Current typical rotation times 3–4 for cop- pice systems or 8–10 years for single stem systems

The RIGES [6] and the LESS-BI scenario [13] have assumed that 75% of the pro-duced organic urban refuse is available for energy use. Furthermore, it is assumed that organic waste production is about 0.3 ton/cap/year, resulting in 3 EJ/year. Des-sus [19] has assumed in his assessment of the biomass energy potential in 2030 that urban waste production could be 0.1–0.3 ton per capita resulting in 1 EJ/year. Hence the range of organic waste could vary from 1 EJ/year to 3 EJ/year.

1.2 Productivity and Availability Constraints

The main challenges concerning productivity and availability of biomass supply are land use in terms of efficiency and availability, good use of primary and co-products, climate change and agricultural management and lifestyle.

1.2.1 Land Use

Efficiency: different sectors – food, feed, fibre, chemicals and energy – compete for land. Therefore in every scenario land should be treated as a competitive parameter and classified according to its physical, chemical and regional characteristics.

The use of low fertility, marginal land has also been modelled in a number of recent studies, indicating that there could be good future potential. However, production in marginal lands has to meet both economic and sustainable criteria in order to become competitive. Currently low-input, low-output plant production is generally not profitable for the farmer and therefore may not provide lower unit cost feedstock for the processor. Moreover, the output is often variable in both quantity and quality. The ultimate products from such a system are likely to carry high unit costs and to limit severely the economic viability of the whole chain. Additionally, because low-input production requires many more acres, the unit impact on the environment is often much greater than from a more intensive system.

Hence, planning efforts should focus on choosing the best available cropping solutions for each region and land type.

Availability: land availability and quality will define the amount and type of feedstocks produced in EU over the coming years. It has been acknowledged in recent studies [20, 21] that increased bioenergy demand can affect both extensive farm areas and grasslands due to potential shifts from existing food and feed production to bioenergy, particularly to lignocellulosic crops. A moderate estimate by the EEA study states that the available land (arable, grassland and olive groves) which could be used for dedicated bioenergy production will increase from 14.7 million ha in 2010 to 25.1 million ha in 2030 [20].

1.2.2 "Good Use" of Both the Primary Resource and Residues

Biomass resources offer substantial variety in terms of chemical and physical properties. So far agricultural and forestry systems operate in such ways that people exploit only part of their production, what is called "primary" product, while they leave unexploited significant "residual" quantities.

The use of residues is expected to maximise the added value of the raw materials and improve the income for the producers (farmers, forest community, etc.).

1.2.3 Climate Change

Climate change is likely to have a significant impact on the availability of biomass as well as on feedstock types produced and their regional distribution. For central and northern Europe, an extension of the growing season in spring and autumn is expected, coupled with higher temperatures during the growing period [20]. This appears to enhance the productivity for both bioenergy crops and forests in these regions. On the other hand, in southern Europe an increased risk of drought could lead to productivity losses and increase the risk of forest fires [20]. Extreme weather conditions can significantly influence the supply of biomass feedstocks and therefore a variety of biomass feedstocks should be supported to secure the viability of the conversion plants.

1.2.4 Agricultural Management and Lifestyle

The efficiency of agricultural lifestyle in terms of inputs, water and resource management, cropping strategies, *etc.* is a key factor which determines availability and long term productivity of land. Improving the respective components according to the climate, ecology and market requirements is essential for future bio-economies.

1.3 Costs

Key issues in biomass feedstock cost analyses are (a) to determine the costs for each feedstock type, (b) to define which components of the supply chain contribute the greatest cost, (c) to estimate cost variability due to regional, sectoral and market demand factors, (d) to compare with their fossil fuel counterparts and (e) to perform sensitivity analysis.

Table 1.4 outlines the major system components in feedstock cost modelling along with the input and output parameters used.

Table 1.4 Supply chain cost system components and modelling parameters

System components	Options	Parameters for modelling	
		Input	Output
Biomass production/ collection	Woodchips, wood wastes	Production costs	Feedstock cost ($€$ GJ^{-1})
resource	Fellings,		
harvest method	Arboricultural cleaning and thinnings	Purchase costs (international trade) harvesting window	
	Energy crops (*e.g.* oil crops, starch crops, sugar crops, SRC, miscanthus, *etc.*)		
	Dry agricultural residues (*e.g.* straw)		
	Wet food and beverage industry residues		
	Other wastes (*e.g.* oils and fats)		
Logistics Pre-treatment	Separating/sorting Mixing/blending Drying Physical state alteration and/or densifying Biochemical/chemical treatment	Equipment capacity Capital costs O& M costs Energy consumption (fuel, energy) Operation time Dry matter losses	Pre-treatment cost per feedstock ($€$ GJ^{-1})
Transport	Truck Ship Rail	Distance Capacity (volume, weight) Fuel consumption Time and costs per load	Transport cost per feedstock ($€$ GJ^{-1})
Storage	Outdoors Indoors	Volume per feedstock type (bales, pellets, *etc.*)	Storage cost per feedstock ($€$ GJ^{-1})

Depending on the feedstock type, some additional characteristics can be further defined. Details are as follows.

Residual Feedstocks

Three types of cost normally apply for residual feedstocks:

- 'Zero' initial cost is when the feedstock is not exploited for any other market and is normally left in the field. However, once a market opportunity arises there would be a cost for the raw material, too.
- 'Negative' costs refer to 'problematic' feedstocks in terms of quality and effective disposal. This normally applies to the 'waste-type' of feedstocks.
- 'Opportunity cost' is the cost the feedstock has in the most important alternative market. Straw prices for animal feed can be an example.

Energy Crops

The main cost elements for energy crops are land rent, establishing/growing, harvesting, storage and transport:

- Land rent: the crop type and management requirements affect the choice of land and its opportunity cost or rent, which in certain cases can range greatly (up to threefold or fourfold for example in cases of irrigated fertile *vs* non-irrigated land).
- Establishing/growing includes such costs as ploughing and harrowing (land preparation), crop establishment (sowing or planting) as well as the cost of seed/stem cuttings, fertiliser and irrigation when required.
- For all the cultivation techniques the costs cover labour, machine depreciation and fuel. Cost estimates for energy crops costs are usually made on a dry ton basis, per cultivated ha or on an energy basis (per GJ).
- Finally, exit costs (grubbing up) for perennials are normally incorporated in the establishment costs.
- Harvesting costs cover the cost of labour and machinery for cutting, chipping and forwarding biomass within the field. Cost of working capital and other costs are also taken into account in the appraisal [15].
- Storage costs: several options are available, the most important being indoor and outdoor storage, on the farm or at a central location. The storage type has implications on the cost and quality of the biomass.
- Transport costs: biomass feedstocks have low bulk densities; with volume and not weight being the limiting factor in their transport [22].
- Another type of cost layout, by breaking total cost by production factor, is also used in cost appraisals for energy crops [15, 23–26]: The key cost factors examined are listed below.
- Labour which is further categorised as skilled and unskilled. The labour required for each stage (establishment, annual) and cultivation technique is calculated.
- Land: land rent is estimated as the opportunity cost of land based on current activity (fallow land, cereals cultivation). Usually this cost of land value is determined by soil productivity combined with economic forces that affect demand for land resources in the under study region.

- Machinery: rent of tractor, harvester and travelling gun is included in the cost analysis.
- Variable inputs: include seed/stem cuttings for crop establishment, fertilisers and pesticides for increased production, irrigation water, *etc.*
- Energy: mainly diesel for machinery operation.

The cost of working capital and overheads are also added to the crop production costs.

Table 1.5 provides cost estimates for different feedstock categories for 2010 with 2020 projections.

Table 1.5 Cost estimates for several feedstock categories in 2010 and 2030

Feedstock	Feedstock production costs (€/GJ)		Sources
	2010	2030	
Residual feedstocks			
Cereal straw	1.1–4.5	1–3	[27, 28]
Wood residues – logs	0.8–0.9	0.8–0.9	Finland [28]
	1–1.5	1–1.5	Estonia [28]
Wood residues – chips	1.6–1.7	1.6–1.7	Finland [28]
	1.7–2.2	1.7–2.2	Sweden [28]
Wood residues – bales	1.2–1.6	1.2–1.6	Finland [28]
	1.3–1.6	1.3–1.6	Sweden [28]
	1.5–3	1.5–2.5	EU27 [28]
Refined wood fuels	1–6	1–4	EU27 [28]
Energy crops			
Oil			
Rapeseed	20	12	[29]
Sunflower	15–18	12	[29]
Sugar			
Sugarbeet	12	8	[29]
Sweet sorghum	3–4	2.5	[15]
Starch			
Wheat	5–11	10	[15]
Barley	5–11	10	[15]
Corn	8–10	10	[15]
Lignocellulosic			
Fiber sorghum	3	2	[25]
Cardoon	3–5	2–3	[23–25, 30]
Miscanthus	4–6	2–3	[23–25, 30]
Giant reed	4–6	2–3	[23–25, 30]
Switchgrass	4–6	2–3	[23–25, 30]
Willow	3–6	2	[29]
Poplar	3–4	2	[29]

1.4 Sustainability

Sustainability involves a set of 'three-dimension' rules/guidelines with respect to economic, environmental and social terms (Figure 1.2) under which biomass fuels should be produced, distributed and used.

As a complex system it can only be properly evaluated in comparison to a given 'reference system' which it will displace in short, medium or long timeframes.

It should be noted here that research on sustainable bioenergy systems is very recent, so that few studies and empirical, field-derived data are available as yet. This applies even more to sustainability issues of bioenergy in developing (mostly Southern) countries, where semi-arid, arid and tropical climates restrict the application of results from "Northern" countries, which have different soils and climates and use different farming systems [31].

Current agricultural practices can have both a negative and positive impact on the environment. For this reason, it is important that any move towards more bioenergy production should aim to support positive development, while at the same time not exacerbating existing pressures on farmland biodiversity and water and soil resources [20].

Source: adapted from BTG [32]

Resource use during agricultural production causes a number of environmental pressures related to soil health and soil quality maintenance, water use and biodiversity.

The main sustainability issues are presented in Table 1.6.

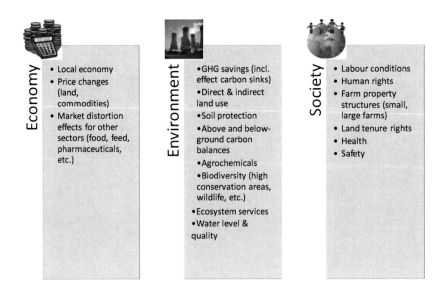

Economy
- Local economy
- Price changes (land, commodities)
- Market distortion effects for other sectors (food, feed, pharmaceuticals, etc.)

Environment
- GHG savings (incl. effect carbon sinks)
- Direct & indirect land use
- Soil protection
- Above and below-ground carbon balances
- Agrochemicals
- Biodiversity (high conservation areas, wildlife, etc.)
- Ecosystem services
- Water level & quality

Society
- Labour conditions
- Human rights
- Farm property structures (small, large farms)
- Land tenure rights
- Health
- Safety

Figure 1.2 Sustainability issues for bioenergy systems

Table 1.6 Sustainability issues

Issue	Impacts	Remarks	References
Land			
Land use change	Land is one of the key 'conflict' factors for the future development of biomass feedstocks. Variability in terms of crop species, cultivation practices, soil and climate conditions makes future predictions and attempts to harmonise modelling assessments extremely difficult. Recent literature provides a general framework to assess the potential for land use to produce biomass feedstocks. As a general rule, the land-use effects of bioenergy-cropping systems must be considered with reference to current land use (if any): • If bioenergy production replaces intensive agriculture, the effects can range from neutral to positive • If it replaces natural ecosystems (forests, wetlands, pasture, *etc.*) the effects will be mostly negative	According to [21] areas to be protected include: High-nature-value areas (*e.g.* intact close-to-nature ecosystems, natural habitats, primary and virgin forests), land needed to maintain critical population levels of species in natural surroundings, and relevant migration corridors must be excluded from bioenergy cropping areas. Adequate buffer zones must be maintained for habitats of rare, threatened or endangered species, as well as for land adjacent to areas needing protection	[21] [20] [33]
Soil			
Soil health and soil quality maintenance	Exploiting crop residues for energy may increase soil erosion and decrease soil organic matter. The fraction of crop residues collectable for biofuel is not easily quantified because it depends on the weather, crop rotation, existing soil fertility, slope of the land and tillage practices. Non-exploited lignocellulosic agricultural residues are either incorporated into the soil or are burned in the field		
Soil erosion	Erosion is a very serious environmental hazard that reduces soil fertility as well as the productivity of agricultural land. Soil erosion remains a significant concern in the EU-15 and appears to be concentrated in the Mediterranean region. It is also an important environmental issue in the new EU Member States	The area of land with a high erosion risk in southern Member States is 22.9 million ha (about 10% of the rural land surface). The extent of this risk extends to one third of land in Portugal, 20% in Greece, 10% in Italy and 1% in France.	[33] [5]

Table 1.6 (continued)

Issue	Impacts	Remarks	References
		A report by ICONA in Spain (1991) shows that some 44% of total land area is affected by erosion and 9 million ha (18%) currently loses more than 50 tons/ha/year, which is considered the critical load for erosion. National average losses of 27 tons/ha/year were reported, compared to soil formation of 2–12 tons/ha/year	
		It is clear that agricultural residues utilization after harvesting the crops in flat fields has no negative effect on soil erosion. Agricultural residues collection and utilization also have no negative effect if the field is planted with autumn sowing crops	
		In contrast, on the hilly, semi-arid and sloping regions, especially of southern EU, maintenance of agricultural residues in the field during the winter months may offer valuable protection against soil erosion if the field is planted with spring sown plants	
Soil nutrients	It is understood that exploitation of agricultural residues for energy purposes will result in negative nutrient balance. During, for example, the combustion process, all the nitrogen contained in the agricultural residues is lost. However, most of the agricultural residues are low in nitrogen content. Residues also contain significant quantities of macronutrients (P, K, Ca, Mg, S) as well as micronutrients. All these nutrients, with the partial exception of sulphur which is released into the atmosphere as SOx, remain in the ashes. If care is taken to return the ashes to the fields, then the nutrient status equilibrium is neutral, with the exception of nitrogen. Therefore, collecting the residues will not be that harmful, taking into account that nutrient losses will be replaced by ashes and nitrogen fertilization in order to maintain soil fertility		[5]

Table 1.6 (continued)

Issue	Impacts	Remarks	References
Biodiversity			
	The following are of particular concern: • conversion from extensive, "high-nature-value" farming to more intensive mono-cropping; • the conversion of primary forests and other habitats into energy plantations, both of which would lead to a severe loss of biodiversity Recent studies from the European Environment Agency (2006, 2007) state that new bioenergy crops (e.g. perennials) and cropping systems (e.g. double cropping, low input agriculture, etc.) can, in certain cases, add to crop diversity and combine a high yield with lower environmental pressures, when compared to intensive food farming systems An environmentally compatible crop mix is recommended based on the regional agro-ecological profile in order to reduce the main environmental pressures of the region, in which bioenergy is produced (EEA, 2006)	According to WWF, 2006, the following are of particular concern: • conversion from extensive, "high-nature-value" farming to more intensive mono-cropping; • the conversion of primary forests27 and other habitats into energy plantations, both of which would lead to a severe loss of biodiversity	[21] [20] [33]
Air quality			
GHG emissions	Bioenergy usually reduces greenhouse gas (GHG) emissions, since its use is carbon-neutral	Since GHG emissions are caused not only by bioenergy cultivation, but also by downstream processing, a GHG standard for bioenergy should address both: • a maximum life-cycle GHG balance of bioenergy cultivation; • the processing of biomass feedstocks which should demonstrate a minimum conversion efficiency (by-products should be taken into account)	[21]
Water			
Water use	Agricultural water use remains a serious concern, especially in south European countries, where water is scarce and highly variable from year to year and where agricultural use of total water consumption is 50% [33].	According to [21], standards should cover both agricultural water use and the protection of water bodies from the impact of agriculture. Suggestions include: Optimized farming systems with low water input	[31] [20] [21]

Table 1.6 (continued)

Issue	Impacts	Remarks	References
	Furthermore, some of the current cropping systems used for biomass supply, like rapeseed and cereals have low efficiency in the use of water and fertilisers but in the future improvements are expected through the use of perennial biomass crops which will have better overall nutrient efficiency than most of the conventional arable crops for biomass production	Water management plans in dry and semi dry regions Time and quantify the application of agrochemicals to maintain the quality and availability of surface and ground water and avoid the negative impacts No untreated sewage water for irrigation Re-use of treated waste-water must be part of the agricultural management system	
Socio-economic impacts			
Macroeconomic	• Rural diversification • Regional growth • Reduced regional trade balance • Export potential	Increased use of crop-based materials and the diversity of feedstocks could improve farming opportunities and respective cash flows as well as secure supply for the related industries. Furthermore, the use of indigenous raw materials assists to retain employment and welfare locally and re-circulate them within the local/regional economy	[34]
Supply	• Increased productivity • Enhanced competitiveness • Labour and population mobility (induced effects) • Improved infrastructure	Supply side effects are the impacts which are the result of improvements in the competitive position of the region, including its attractiveness to inward investment. These effects are regionally/locally specific and relate to changes and improvements in regional productivity, enhanced competitiveness, as well as any investment in resources to accommodate any inward migration that may result from the development	[34]
Demand	• Employment • Income and wealth creation • Induced investment • Support of related industrie	Demand side effects are primarily quoted in terms of employment and regional income	

1.5 Challenges/Recommendations

The main challenges that the biomass supply sector is facing are:

- Supply diverse markets and consumer needs (bio-cascading solutions). As demand for resources is increasing due to policy targets and industrial requirements for ecological resource based products it is essential to develop supply solutions that can provide the base for a range of industries (chemicals, pharmaceuticals, building/construction, energy and fuels) and satisfy their end product requirements.
- Expand feedstock supplies including sustainable trade. Currently, regarding commercial applications, the biomass feedstock matrix is comprised of a restricted number of biomass types like oilseeds for biodiesel, corn, cereals and sugarbeet for bioethanol, manure and sludges for biogas as well as wood and straw for stationary CHP power plants and small scale heat applications. The need to expand the feedstock matrix is strong and requires a harmonised sustainability criteria scheme that can apply to all the resources and cover their wide regional distribution.

Table 1.7 presents a SWOT analysis for the supply of solid biomass in EU27.

- Meet the quality requirements of the processes through improved certification. As the end use markets vary, appropriate certification schemes should be developed and applied to ensure quality throughout production process and end product specifications. Interactions and links among the several schemes should be strengthened appropriately.
- Maximize yield per unit area while minimizing negative environmental impacts. Research should focus on existing and improved crops, adaptation to local conditions, improved rotation, land management, "green biotechnology" (genetic engineering, assisted breeding, *etc.*).

Recommendations for future solid biomass supply (Figure 1.3) include:
Land:

- Create land strategies suitable for both the soil-climatic characteristics and the environmental and socio-economic conditions prevailing in the region. The strategies should not be competitive but they must be complementary under a set of economic and sustainable criteria.

Feedstocks:

- Promote the use of both the primary and residual forms of agricultural and forestry operations under sustainable rules that take into account soil properties and balance of the ecosystem.
- Integrated approaches for crops and by-products require in some cases that the residues from forestry, agriculture and similar sectors which are appropriate for energy applications should be considered as fuels and not as wastes.
- Develop new high-yield and low-input agricultural systems with breeding of crops and trees with optimised characteristics to match different markets.

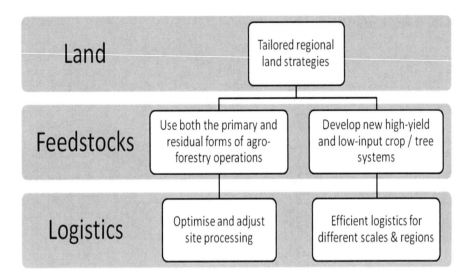

Figure 1.3 Roadmap for future biomass supply

Table 1.7 SWOT analysis for solid biomass supply

Strengths	Weaknesses
• There is a highly competent RTD background in EU27 comprising both of human resources and the respective research infrastructures, methodologies and tools • Good partnerships within and outside EU borders exist and provide the basis for future research work, transfer of knowledge and technology, exchange of scientists, *etc.* (EU-USA, EU-LA, EU-China, *etc.*) • The required critical mass is there and string collaborations exist in the research and industrial level • Demand is getting stronger for secure and sustainable biomass supply	• There is a complex matrix of feedstocks with different characteristics and logistic/handling requirements • Seasonality (harvest window). Biomass availability is hindered by short harvest window (harvest window is not critical if storage of biomass is possible, which is the case for most systems). Therefore appropriate strategies should be created to avoid disruption in the supply • Large volume handling/ logistics is required to develop industrial scale of biofuels. So far the feedstock management systems are designed to meet small-medium scale requirements. An upgrade is considered essential
Opportunities	Threats
• Favourable political floor – The Biofuels Directive (2003/30/EC) – The Biomass Action Plan (COM(2005) 628)	• Time: science development/implementation needs to speed up • Myths: biotechnology and GM products • Sustainability: environmental impacts
– A Strategy for Biofuels >> (COM(2006) 34) – 120 KB PDF – Priority is given to biofuels research in the Seventh RTD Framework Programme (FP7) • High oil prices enhance the competitive position of biomass and biofuels in the market • There is an increasing industrial interest in the field with substantial investment going on	• Links: interfaces to target multi-functionality (food, fibre, fuel, feed)

Logistics:

- Site processing of agricultural, forestry products needs further optimization and adjustment so that the physical form and the quality characteristics of the material meet the conversion technology requirements.
- Develop efficient biomass logistic systems (harvesting/collection/storage) for different conversion concepts at different scales.
- Biomass trade should be regulated not only with quality and safety protocols but also with sustainability standards.

1.6 Conclusions

The diversity of landscapes and related economic activities in Europe provides a wide range of potential feedstocks to supply the future bio-economy. In order to secure year-round sustainable supply for the different end use markets, biomass production should be linked as an adaptation strategy to climate change by tackling key issues such as water management, rising temperatures, soil erosion, *etc.*

Sustainable land strategies should be created covering the above factors along with being compatible with the climatic, environmental and socio-economic profiles in each region.

References

1. Eurob'server: The state of renewable energies in Europe, Edition 2008
2. Berndes G, Hoogwijk M, van den Broek R (2003) The contribution of biomass in the future global energy supply: a review of 17 studies. Biomass and Energy 25:1–28
3. Fischer G, Schrattenholzer L (2001) Global bioenergy potentials through 2050. Biomass Bioenerg 20:151–159
4. IEA, OECD (2003) OECD balances of non-OECD countries 1999–2000, IEA/OECD, Paris
5. Dalianis C, Panoutsou C (1995) Energy potential of agricultural residues in EU. EUREC Network on Biomass (Bioelectricity) Final Report. Contract No: RENA CT 94-0053
6. Johansson TB, Kelly H, Reddy AKN, Williams RH (1993) Renewable fuels and electricity for a growing world economy – defining and achieving the potential. In: Johansson TB, Kelly H, Reddy AKN, Williams RH (eds) Renewable energy: sources for fuels and electricity, Island Press, Washington, DC
7. Kaltschmitt M, Neubarth J (2000) Biomass for energy – an option for covering the energy demand and contributing to the reduction of GHG emissions? Workshop Proceedings, Workshop on Integrating Biomass Energy with Agriculture, Forestry and Climate Change Policies in Europe, December 2000, Imperial College, London, UK
8. Kaltschmitt M, Dinkelbach L (1997) Biomass for energy in Europe – status and prospects In: Kaltschmitt M, Bridgwater AV (eds) Biomass gasification and pyrolysis – state of the art and future prospects, CPL Scientific, Newbury, UK
9. Lazarus M (1993) Towards a fossil free energy future. Boston: Stockholm Environment Institute 240

10. Hall DO, Rosilo-Calle F, Williams RH, Woods J (1993) Biomass for energy: supply prospects. In: Johansson TB, Kelly H, Reddy AKN, Williams RH (eds) Renewable energy: sources for fuels and electricity, Island Press, Washington, DC
11. Parikka M (2004) Global biomass fuel resources. Biomass Bioenerg 27:613–620
12. Wirsenius S (2000) Human use of land and organic materials – modelling the turnover of biomass in the global food system. Chalmers University 255
13. Swisher J, Wilson D (1993) Renewable energy potentials. Energy 18(5):437–459
14. FAO (2006) State of the world's forests – 2006. www.fao.org
15. Panoutsou C, Namatov I, Lychnaras V, Nikolaou A (2005) Biofuels in Greece: production and environmental impacts. In Greek. RENES conference 23–25 February, Athens, Greece
16. FAO (2002) FAOSTAT-database 2002. http://ww.fao.org
17. FAO (1993) Energy conservation in the mechanical forest industries. Forestry Paper 93
18. Cosentino LS (2007) Personal communication
19. Dessus B, Devin B, Pharabod F (1992) World potential of renewable energies, CNRS-PIRSEM Paris 70
20. EEA (2006) European Environmental Agency. How much bioenergy can Europe produce without harming the environment?
21. WWF (2006) Sustainability standards for bioenergy, Authors: Fritsche UR, Hünecke K, Hermann A, Schulze F, Wiegmann K, Berlin, November 2006, Imke Lübbeke, WWF Germany
22. Royal Commission on Environmental Pollution (2004) Biomass as a renewable energy source. http://www.rcep.org.uk/bioreport.htm
23. Lychnaras V, Rozakis S, Soldatos P, Tsiboukas K, Panoutsou C (2007) Economic Analysis of Perennial Energy Crops Production in Greece under the Current CAP. Proceedings of the 15th European Biomass Conference and Exhibition 7–11 May 2007 ICC Berlin Germany
24. Lychnaras V, Rozakis S, Soldatos P, Tsiboukas K, Panoutsou C (2007) Economic Analysis of Perennial Energy Crops Production in Greece under the Current CAP. Proceedings of the 15th European Biomass Conference and Exhibition 7–11 May 2007, ICC, Berlin, Germany
25. Monti A, Fazio S, Lychnaras V, Soldatos P, Venturi G (2007) A full economic analysis of switchgrass under different scenarios in Italy estimated by BEE model, Biomass Bioenerg, 31(4):177–185
26. Soldatos P, Lychnaras V, Asimakis D, Christou M (2004) Bee – Biomass Economic Evaluation: A Model for the Economic Analysis of Biomass Cultivation, 2nd World Conference and Technology Exhibition on Biomass for Energy, Industry and Climate Protection, 10–14 May 2004, Rome, Italy
27. Allen J, Browne M, Hunter A, Boyd J, Palmer H (1996) Logistics management and costs of biomass fuel supply. Int J Phys Distrib 28(6):463–477
28. Siemons R, Vis M, Berg D, Chesney I, Whiteley M, Nikolaou A (2004) Bioenergy's role in the EU energy market: a view of developments until 2020. Report to the European Commission
29. Faaij A (2006) Bioenergy in EU changing technology. Energ Policy 34(3):322–342
30. Christou M, Lychnaras V, Bookis I, Kontopoulos G, Panoutsou C, Alexopoulou E (2009) Could energy crops be an economic and sustainable option for heat and electricity? In 17th European Biomass Conference, Hamburg
31. JRC (2007) Proceedings of the expert consultation: short rotation forestry, short rotation coppice and perennial grasses in the European Union: agro-environmental aspects, present use and perspectives. In: Dallemand JF, Petersen JE, Karp A (eds) Short rotation forestry, short rotation coppice and perennial grasses in the European Union. Harpenden, United Kingdom: (European Commission Joint Research Centre)
32. BTG. Sustainability criteria and certification systems for biomass production: a report for DG TREN – European Commission. Biomass Technology Group; 2008
33. EEA (2007) European Environment Agency technical report #12/2007: estimating the environmentally compatible bioenergy potential from agriculture. European Environment Agency
34. Madlener R, Myles H (2000) Modelling socio-economic aspects of bioenergy systems: a survey prepared for IEA Bioenergy Task 29

Chapter 2
European Standards for Fuel Specification and Classes of Solid Biofuels

Eija Alakangas

Abstract The technical committee developing the draft standard to describe all forms of solid biofuels within Europe (CEN/TC 335) has published 27 technical specifications for solid biofuels. The two most important are classification and specification (CEN/TS 14961) and quality assurance (CEN/TS 15234). Now these technical specifications are upgraded to full European standards (EN). Both these standards will be published as multipart standards. Part 1 – General requirements of EN 14961-1 includes all solid biofuels and is targeted for all user groups. The classification of solid biofuels is based on their origin and source and biofuels are divided to four sub-categories: (1) Woody biomass, (2) Herbaceous biomass, (3) Fruit biomass, and (4) Blends and mixtures. The quality tables were prepared only for major traded forms. Parts 2–6 are product standards, which are targeted for non-industrial use. Non-industrial use means fuel intended to be used in smaller appliances, such as in households and small commercial and public sector buildings. In the product standards all properties are normative and they are bound together to form a class, for example A1, A2, and B. Although these product standards may be obtained separately, it should be recognized that they require an understanding of the standards based on and supporting EN 14961-1. This chapter concentrates on Part 1 of EN 14961, which was published in 2010. The remaining five product standards are being drafted and are at the voting stage, with an expected publication date within 2010.

Acknowledgments to European commission funding the projects, which supported the standardization work and also members of working group 2 in CEN/TC 335, which have actively participated in drafting standards and collecting information.

E. Alakangas (✉)
VTT, Technical Research Centre of Finland, P.O. Box 1603,
FI-40101 Jyväskylä, Finland
e-mail: eija.alakangas@vtt.fi

2.1 Introduction

The European Committee for Standardization, CEN under committee TC335 has published 27 technical specifications (pre-standards) for solid biofuels. Now these technical specifications are upgraded to full European standards (EN).When EN-standards are in force the national standards in Europe have to be withdrawn or adapted to these EN-standards. The two most important technical specifications being developed deal with classification and specification (EN 14961 [14]) and quality assurance for solid biofuels (EN 15234 [23]). Both these standards will be published as multipart standards. Part 1 – General requirements of EN 14961-1 includes all solid biofuels and is targeted for all user groups [1, 6–10].

This European Standard determines the fuel quality classes and specifications for solid biofuels. The scope of the CEN/TC 335 only includes solid biofuels originating from the following sources:

- products from agriculture and forestry;
- vegetable waste from agriculture and forestry;
- vegetable waste from the food processing industry;
- wood waste, with the exception of wood waste which may contain halogenated organic compounds or heavy metals as a result of treatment with wood pre-servatives or coating, and which includes in particular such wood waste origi-nated from construction and demolition waste;
- fibrous vegetable waste from virgin pulp production and from production of paper from pulp, if it is co-incinerated at the place of production and heat gen-erated is recovered;
- cork waste.

Note 1   To avoid any doubt, demolition wood is not included in the scope of this European Standard. Demolition wood is "used wood arising from demolition of buildings or civil engineering installations" (EN14588[17]).
Note 2   Aquatic biomass is not included in the scope of EN 14961-1 [14].

Development of standards has been supported by several projects. EUBIO-NET II [5] has collected experiences of the CEN solid biofuels standards from the market workers during 2006. A selected group of fuel market workers (47) in different countries were interviewed to decide on the concept of the functionality of the EN 14961. The level at which the standards are used or will be used in eve-ryday fuel trade was studied, and the experienced advantages and disadvantages were collected. In the BioNorm II project 25 partners from 11 European countries tested different versions of EN 14961-1 in specifying their solid biofuels accord-ing to standards [4]. Companies involved in the testing were producing pellets, briquettes, wood chips, and hog fuel from woody biomass. Olive residues and reed canary grass bales were also within specified fuels. These experiments have helped in setting threshold values for solid biomass fuels and also drafting prop-erty tables for new traded forms, *e.g.*, bales from herbaceous biomass, olive resi-dues, and energy grain.

2.2 Classification of Biomass Sources

The classification of solid biofuels is based on their origin and source. The fuel production chain of fuels shall be unambiguously traceable back over the whole chain.

The solid biofuels are divided into the following sub-categories for classification in EN 14961-1 [14]:

1. woody biomass (Tables 2.1 and 2.2);
2. herbaceous biomass (Tables 2.3 and 2.4);
3. fruit biomass (Tables 2.5 and 2.6);
4. blends and mixtures.

The purpose of classification is to allow the possibility to differentiate and specify raw material based on origin with as much detail as needed. The quality classification in a table form was only prepared for major traded solid biofuels.

Table 2.1 Classification of 1.1 Forest, plantation and other virgin wood in EN 14961-1 [14]

1.1.1 Whole trees without roots	1.1.1.1 Broadleaf
	1.1.1.2 Coniferous
	1.1.1.3 Short rotation coppice
	1.1.1.4 Bushes
	1.1.1.5 Blends and mixtures
1.1.2 Whole trees with roots	1.1.2.1 Broadleaf
	1.1.2.2 Coniferous
	1.1.2.3 Short rotation coppice
	1.1.2.4 Bushes
	1.1.2.5 Blends and mixtures
1.1.3 Stemwood	1.1.3.1 Broadleaf
	1.1.3.2 Coniferous
	1.1.3.3 Blends and mixtures
1.1.4 Logging residues	1.1.4.1 Fresh/Green, Broadleaf (including leaves)
	1.1.4.2 Fresh/Green, Coniferous (including needles)
	1.1.4.3 Stored, broadleaf
	1.1.4.4 Stored, coniferous
	1.1.4.5 Blends and mixtures
1.1.5 Stumps/roots	1.1.5.1 Broadleaf
	1.1.5.2 Coniferous
	1.1.5.3 Short rotation coppice
	1.1.5.4 Bushes
	1.1.5.5 Blends and mixtures
1.1.6 Bark (from forestry operations)[a]	
1.1.7 Segregated wood (Figure 2.1) from gardens, parks, roadside maintenance, vineyards, and fruit orchards	
1.1.8 Blends and mixtures	

[a] Also includes cork

Table 2.2 Classification of 1.2 By-products and residues from wood processing industry and 1.3 Used wood in EN 14961-1 [14]

1.2 By-products and residues from wood processing industry	
1.2.1 Chemically untreated wood residues	1.2.1.1 Without bark, broadleaf
	1.2.1.2 Without bark, coniferous
	1.2.1.3 With bark, broadleaf
	1.2.1.4 With bark, coniferous
	1.2.1.5 Bark (from industry operations)[a]
1.2.2 Chemically treated wood residues (Figure 2.2), fibers and wood constituents	1.2.2.1 Without bark
	1.2.2.2 With bark
	1.2.2.3 Bark (from industry operations)[a]
	1.2.2.4 Fibers and wood constituents
1.2.3 Blends and mixtures	
1.3 Used wood	
1.3.1 Chemically untreated wood	1.3.1.1 Without bark
	1.3.1.2 With bark
	1.3.1.3 Bark[a]
1.3.2 Chemically treated wood	1.3.2.1 Without bark
	1.3.2.2 With bark
	1.3.2.3 Bark[a]
1.3.3 Blends and mixtures	

[a] Also includes cork

Note 1 If appropriate, the actual species (*e.g.*, spruce, wheat) of biomass can also be stated. Wood species can be stated, *e.g.*, according to EN 13556 Round and sawn timber Nomenclature [11].

Note 2 Chemical treatment before harvesting of biomass does not need to be stated. Where any operator in the fuel supply chain has reason to suspects serious contamination of land (*e.g.*, coal slag heaps) or if planting has been used specifically for the sequestration of chemicals or biomass is fertilized by sewage sludge (issued from waste water treatment or chemical process), fuel analysis should be carried out to identify chemical impurities such as halogenated organic compounds or heavy metals.

The EN14961-1 also includes wood waste if it does not contain halogenated organic compounds or heavy metals as a result of treatment with wood preservatives or coating. The EU-funded BioNormII project clarified which fractions of wood waste can be defined as solid biofuel [2, 27].

In addition to virgin wood, solid biofuels derived from the by-products and residues of the wood processing industry, as well as post-society used wood, are also part of woody biomass (Figure 2.3). Part of the woody material under the heading "used wood" (class 1.3) can justifiably be classified as biomass. Due to the absence of clear guidelines and definitions, the classification of used wood into either waste or biomass remains debatable in the case of certain fractions of wood residues and wastes. Classes A, B, C, and D for used wood and industrial wood residues and by-products were proposed as a result of the study. Wood waste in

Table 2.3 Classification of 2.1 Herbaceous biomass from agriculture and horticulture [14]

2.1.1 Cereal crops	2.1.1.1 Whole plant
	2.1.1.2 Straw parts
	2.1.1.3 Grains or seeds
	2.1.1.4 Husks or shells
	2.1.1.5 Blends and mixtures
2.1.2 Grasses	2.1.2.1 Whole plant
	2.1.2.2 Straw parts
	2.1.2.3 Seeds
	2.1.2.4 Shells
	2.1.2.5 Blends and mixtures
2.1.3 Oil seed crops	2.1.3.1 Whole plant
	2.1.3.2 Stalks and leaves
	2.1.3.3 Seeds
	2.1.3.4 Husks or shells
	2.1.3.5 Blends and mixtures
2.1.4 Root crops	2.1.4.1 Whole plant
	2.1.4.2 Stalks and leaves
	2.1.4.3 Root
	2.1.4.4 Blends and mixtures
2.1.5 Legume crops	2.1.5.1 Whole plant
	2.1.5.2 Stalks and leaves
	2.1.5.3 Fruit
	2.1.5.4 Pods
	2.1.5.5 Blends and mixtures
2.1.6 Flowers	2.1.6.1 Whole plant
	2.1.6.2 Stalks and leaves
	2.1.6.3 Seeds
	2.1.6.4 Blends and mixtures
2.1.7 Segregated herbaceous biomass from gardens, parks, roadside maintenance, vineyards, and fruit orchards	
2.1.8 Blends and mixtures	

classes A and B is solid biofuel as defined, with given restrictions (do not contain halogenated organic compounds or heavy metals as a result of treatment with wood preservatives or coating). Wood waste in class C falls under the Waste Incineration Directive (WID) 2000/76/EC, and is solid recovered fuel. Wood waste in class D includes preservatives and shall be disposed of according to the Hazardous Waste Incineration Directive (94/67/EC).

Figure 2.1   Segregated wood from road maintenance

Figure 2.2   Chemically treated wood residues from process industry, which do not contain heavy metals or halogenated organic compounds

Table 2.4 Classification of 2.2 By-products and residues from herbaceous processing industry [14]

2.2.1 Chemically untreated herbaceous residues	2.2.1.1 Cereal crops and grasses
	2.2.1.2 Oil seed crops
	2.2.1.3 Root crops
	2.2.1.4 Legume crops
	2.2.1.5 Flowers
	2.2.1.6 Blends and mixtures
2.2.2 Chemically treated herbaceous residues	2.2.2.1 Cereal crops and grasses
	2.2.2.2 Oil seed crops
	2.2.2.3 Root crops
	2.2.2.4 Legume crops
	2.2.2.5 Flowers
	2.2.2.6 Blends and mixtures
2.2.3 Blends and mixtures	

Group 2.2 also includes residues and by-products from the food processing industry.

Table 2.5 Classification of 3.1 Fruit biomass [14]

3.1 Orchard and horticulture fruit	3.1.1 Berries	3.1.1.1 Whole berries
		3.1.1.2 Flesh
		3.1.1.3 Seeds
		3.1.1.4 Blends and mixtures
	3.1.2 Stone/kernel fruits	3.1.2.1 Whole fruit
		3.1.2.2 Flesh
		3.1.2.3 Stone/kernel
		3.1.2.4 Blends and mixtures
	3.1.3 Nuts and acorns	3.1.3.1 Whole nuts
		3.1.3.2 Shells/husks
		3.1.3.3 Kernels
		3.1.3.4 Blends and mixtures
	3.1.4 Blends and mixtures	

Table 2.6 Classification of 3.2 By-products and residues from fruit processing industry, 3.3 Blends and mixtures of fruit, and 4 Blends and mixtures [14]

3.2 By-products and residues from fruit processing industry	3.2.1 Chemically untreated fruit residues	3.2.1.1 Berries
		3.2.1.2 Stone/kernel fruits
		3.2.1.3 Nuts and acorns
		3.2.1.4 Crude olive cake (Figure 2.4)
		3.2.1.5 Blends and mixtures
	3.2.2 Chemically treated fruit residues	3.2.2.1 Berries
		3.2.2.2 Stone/kernel fruits
		3.2.2.3 Nuts and acorns
		3.2.2.4 Exhausted olive cake
		3.2.2.5 Blends and mixtures
	3.2.3 Blends and mixtures	
3.3 Blends and mixtures		

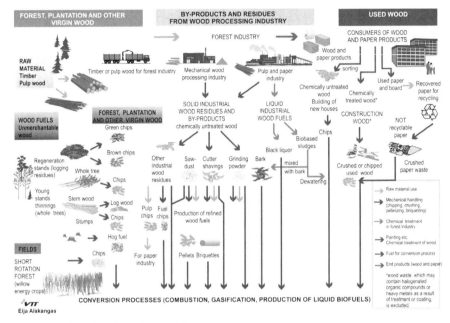

Figure 2.3   Classification of woody biomass

Figure 2.4   Olive residues

2.3 Fuel Specification and Classes – Multipart Standard

EN 14961 consists of the following parts, under the general title Solid biofuel – Fuel specification and classes [14]:

Part 1: General requirements (final draft) [14]

Part 2: Reed canary grass (Figure 2.5) for non-industrial use (under development) [18]

Part 3: Wood briquettes for non-industrial use (under development) [19]

Part 4: Wood chips for non-industrial use (under development) [20]

Part 5: Firewood for non-industrial use (under development) [21]

Part 6: Non-woody pellets for non-industrial use (under development) [22]

Properties to be specified are listed in Tables 3–14 of EN 14961-1 for the following traded forms of solid biofuels: (EN 14961-1) briquettes, pellets, wood chips, hog fuel, log wood/firewood, sawdust, shavings, bark, straw bales, reed canary grass (Figure 2.6) bales and Miscanthus bales, energy grain, olive residues, and fruit seeds. A general master table (Table 15 in EN 14961-1) is to be used for solid biofuels not covered by Tables 3–14. [14]. In Appendix 1 the classification table for wood pellets and in Appendix 2 for wood chips is presented.

The classification is flexible in EN 14961-1, and hence the producer or the consumer may select from each property class the classification that corresponds to the produced or desired fuel quality. This so-called "free classification" in Part 1 does not bind different characteristics with each other. An advantage of this classification is that the producer and the consumer may agree upon characteristics case-by-case.

The most significant characteristics are mandatory (= normative) and shall be given in the fuel specification EN 14961-1. These characteristics vary for different traded forms, while the most significant characteristics for all solid biofuels are moisture content (M), particle size/dimensions (P or D/L), and ash content (A). For example, the average moisture content of fuels is given as a value after the symbol (*e.g.*, M10), which means that the average moisture content of the fuel shall be ≤10 wt%. Some characteristics, *e.g.*, bulk density (BD), are voluntary, informative (see appendices 1 and 2 for pellets and wood chips respectively).

In these product standards, non-industrial use means fuel intended to be used in smaller appliances, such as in households and small commercial and public sector buildings. Although these product standards may be obtained separately, it should be recognized that they require an understanding of the standards based on and supporting EN 14961-1. The product standards will be drafted and they will be ready for voting at the end of 2009.

In the product standards all properties are normative and they are bound together to form a class, for example A1, A2, and B for wood pellets. Property class A1 for wood pellets represents virgin woods and chemically untreated wood residues low in ash and chlorine content. Fuels with slightly higher ash content and/or chlorine content fall within grade A2. In property class B chemically treated industrial wood by-products and residues and used wood are allowed (Appendix 2) if threshold values for minor elements are fulfilled.

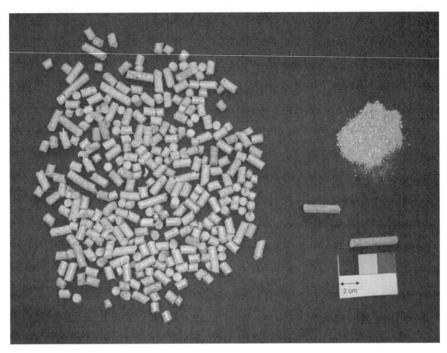

Figure 2.5   Wood pellets classified according to EN 14961-1

Figure 2.6   Reed canary grass. Photo Vapo Oy

To support the development of product standards BioNormII project carried out combustion tests [3, 15, 16, 24–26] with small-scale wood burning appliances to find out if fuel properties (moisture, particle size, different biomass species, and ash content) affected the combustion, boiler efficiency, and emission formation. Tests were carried out by using standard EN303-5 (boilers [12]) and EN15250 (stove [13]). The appliances used were a conventional and modern log wood boiler, wood chips boiler, conventional and modern pellet boiler, and conventional heat retaining stove. The most important fuel properties were changed in tests, *e.g.*, moisture content, particle size, and ash content. Different wood species were used for log wood [16, 24] and different fruit biomass was tested in a conventional wood chip boiler [15]. Particle size of log wood did not have a big influence on emissions and boiler efficiency when a modern log wood boiler was used, but it had an effect on stoves. It was recommended that the diameter for stoves should be less than 15 cm. Based on the combustion tests the moisture content for log wood should be less than 20 wt%, but not lower than 10 wt%. Wood and agrobiomass pellets (diameters 6 mm and 8 mm) were used with different ash contents [26]. If ash content is higher than 0.5 wt% on a dry basis, particle size emissions increased. Ash content of woody biomass is usually low, but some wood species have higher ash content than 0.5 wt%.

Emissions were higher for different agrobiomass pellets and residues than for wood fuels [26]. Appliances used in tests were not designed for agrobiomass so this also affected the results. CEN/TC 335 is also developing a product standard for non-woody pellets.

To protect the small-scale consumer some minor elements are normative for wood pellets [18] and briquettes [19]. For wood chips [20] minor elements are normative if wood chips are produced from short rotation forestry and used wood.

If the properties being specified are sufficiently known through information about the origin and handling (or preparation method combined with experience) then physical/chemical analysis may not be needed.

To ensure resources are used appropriately and the declaration is accurate, utilize the most appropriate measure from those below:

1. using typical values, *e.g.*, laid down in Annex B in EN 14961-1, or obtained by experience;
2. calculation of properties, *e.g.*, by using typical values and considering documented specific values;
3. carrying out of analysis: (a) with simplified methods if available, (b) with reference methods.

2.4 Examples of Fuel Specification

The quality management system in ISO 9001 generally consists of quality planning, quality control, quality assurance, and quality improvement. The EN 15234 [23] covers fuel quality assurance and quality control. It covers quality assurance of the supply chain and information to be used in quality control of the product, so that traceability exists and confidence is given by demonstrating that all processes along the overall supply chain of solid biofuels up to the point of the delivery to the end-user are under control.

Quality assurance aims to provide confidence that a steady quality is continually achieved in accordance with customer requirements.

The methodology shall allow producers and suppliers of solid biofuels to design a fuel quality assurance system to ensure that:

- traceability exists;
- requirements that influence the product quality is controlled;
- the end-user can have confidence in the product quality.

A fuel quality declaration for the solid biofuel shall be issued by the supplier to the end-user or retailer. The fuel quality declaration shall be issued for each defined lot. The quantity of the lot shall be defined in the delivery agreement. The supplier shall date the declaration and keep the records for a minimum of 1 year after the delivery. The fuel quality declaration shall state the quality in accordance with EN 14961 (see example in Figure 2.7).

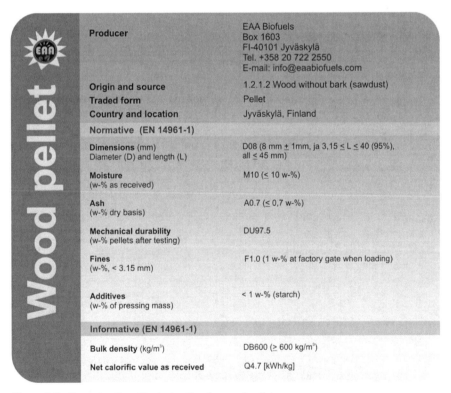

Figure 2.7 Example of quality declaration for wood pellets

Table 2.7   Specification of fuel properties according to EN 14961-1 for wood chips

	Wood chips – EN 14961-1	
Normative	Origin	1.1.1.1 Whole trees without roots (broadleaf)
	Particle size, P (mm)	P45A
	Ash, A (wt% of dry matter)	A1.5
Informative	Bulk density (BD) as received, kg/m³ (loose)	BD250

The fuel quality declaration shall as a minimum include:

- supplier (body or enterprise) including contact information;
- a reference to EN 15234 – Fuel quality assurance;
- origin and source (according to appropriate part of EN 14961;
- country and location where the biomass is harvested or first traded as biofuel;
- traded form (*e.g.*, pellet);
- normative properties;
- chemical treatment if chemically treated biomass is traded;
- signature (by operational title or responsibility), name, date, and place.

The fuel quality declaration can be approved electronically. Signature and date can be approved by signing of the waybill or stamping of the packages in accordance with the appropriate part of EN 14961.

In Table 2.7 and Figure 2.8 there are examples of specification of wood chips and wood pellets.

Figure 2.8   Wood chips (*left*) and hog fuel (*right*)

2.5 Summary

CEN/TC 335 of Solid biofuels has published 27 technical specifications for solid
biofuels and these are upgraded to EN-standards. The two most important are Fuel
classification and specification [14] and Fuel quality assurance [23]. The classification of solid biofuels is based on their origin and source and biofuels are divided
to four sub-categories: (1) Woody biomass, (2) Herbaceous biomass, (3) Fruit
biomass, and (4) Blends and mixtures. The quality tables were prepared only for
major traded forms. The classification is flexible, and this "free classification"
does not bind different characteristics with each other.

The upgrading of the solid biofuels technical specifications is also supported by
EU-funded projects. EUBIONET II has collected feedback from 47 market workers in Europe [4]. The FP6 project BioNorm II [1, 3] is carrying out pre-normative
research for all technical specifications. Fuel classification and classes standards
have been tested in BioNormII project by ten companies. Also, combustion tests
[3, 15, 24, 25, 26] were carried out to support setting threshold values especially
for product standards. The Phydades-project (www.phydades.info) is collecting
property information of solid biofuels for Biodat-database and training laboratory
staff for fuel analysis based on CEN methods.

Traders and fuel suppliers of pellets, wood chips, and hog fuel (Figure 2.8) to
district heating and power stations have found the free classification system practical according to the studies of EUBIONET II [4]. Quality declaration is also used
in pellet packages. More quality categories, in which properties are bound together
and form a class, are used in conveyance from log wood traders and retailers to
domestic consumers. The Committee CEN/TC 335 made a decision to prepare
product standards for non-industrial use.

Part 1 of EN 14961 was published in 2010. The remaining five product standards are being drafted and are at the voting stage, with an expected publication
date within 2010.

Appendix 1. Specification of Properties for Pellets (EN 14961-1) [14]

Table 2.8 Specification of properties for wood pellets

Normative

Master table	
Origin: According to Table 1 of EN 14961-1	Woody biomass (1), Herbaceous biomass (2), Fruit biomass (3), Blends and mixtures (4)
Traded Form	Pellets
Dimensions (mm)	Diameter (D) and Length (L)[a]
D06	6 mm ± 1.0 mm and 3.15 ≤ L ≤ 40 mm
D08	8 mm ± 1.0 mm, and 3.15 ≤ L ≤ 40 mm
D10	10 mm ± 1.0 mm, and 3.15 ≤ L ≤ 40 mm
D12	12 mm ± 1.0 mm, and 3.15 ≤ L ≤ 50 mm
D25	25 mm ± 1.0 mm, and 10 ≤ L ≤ 50 mm
Moisture, M (wt% as received)	Method: EN14774
M10	≤ 10 %
M15	≤ 15 %
Ash, A (wt% of dry basis)	Method: EN 14775
A0.5	≤ 0.5%
A0.7	≤ 0.7%
A1.0	≤ 1.0%
A1.5	≤ 1.5%
A2.0	≤ 2.0%
A3.0	≤ 3.0%
A5.0	≤ 5.0%
A7.0	≤ 7.0%
A10.0	≤ 10.0%
A10.0+	> 10.0%
Mechanical durability, DU (wt% of pellets after testing) Method: EN15210-1	
DU97.5	≥ 97.5%
DU96.5	≥ 96.5%
DU95.0	≥ 95.0%
DU95.0–	< 95.0% (minimum value to be stated)
Amount of fines, F (wt%, < 3.15 mm) after production when loaded or packed[b] (EN 15149-1)	
F1.0	≤ 1.0%
F2.0	≤ 2.0%
F3.0	≤ 3.0%
F5.0	≤ 5.0%
F5.0+	> 5.0% (maximum value to be stated)
Additives (wt% of pressing mass)	
Type and content of pressing aids, slagging inhibitors or any other additives have to be stated	
Bulk density (BD) as received (kg/m^3) (Method: EN 15103)	
BD550	≥ 550 kg/m^3
BD600	≥ 600 kg/m^3
BD650	≥ 650 kg/m^3
BD700	≥ 700 kg/m^3
BD700+	> 700 kg/m^3 (minimum value to be stated)
Net calorific value as received, Q (MJ/kg or kWh/kg) (Method: EN 14918)	
Minimum value to be stated	

Specification of properties for wood pellets (*continued*)

	Sulfur, S (wt% of dry basis) (Method: EN15289)	
	S0.02	≤0.02%
	S0.05	≤0.05%
	S0.08	≤0.08%
	S0.10	≤0.10%
	S0.20	≤0.20%
	S0.20+	>0.20% (maximum value to be stated)
Informative/Normative[d]	**Nitrogen, N** (wt% of dry basis) (Method: EN 15104)	
	N0.3	≤0.3%
	N0.5	≤0.5%
	N1.0	≤1.0%
	N2.0	≤2.0%
	N3.0	≤3.0%
	N3.0+	>3.0% (maximum value to be stated)
Normative/Informative[d]	**Chlorine, Cl** (wt% of dry basis) (Method: EN15289)	
	Cl0.02	≤0.02%
	Cl0.03	≤0.03%
	Cl0.07	≤0.07%
	Cl0.10	≤0.10%
	Cl0.10+	>0.10% (maximum value to be stated)
Informative	**Ash melting behavior** (°C)	Deformation temperature, DT should be stated (method: EN 15370-1)

[a] Amount of pellets longer than 40 (or 50 mm) can be 5 wt%. Maximum length for classes D06, D08 and D10 shall be <45 mm

[b] Fines shall be determined by EN 15149-1

[c] The maximum amount of additive is 20 wt% of pressing mass. Type stated (*e.g.*, starch). If amount is greater, then raw material for pellet is blend

[d] Sulfur, nitrogen and chlorine are normative for the following biomass:
Chemically treated biomass (1.2.2; 1.3.2; 2.2.2; 3.2.2) or if sulfur containing additives have been used. Sulfur, nitrogen, and chlorine are informative for all fuels that are not chemically treated (see the exceptions above)

Note Special attention should be paid to the ash melting behavior for some biomass fuels, for example eucalyptus, poplar, short rotation coppice, straw, miscanthus, and olive stone

Appendix 2. Specification of Properties for Wood Chips (EN 14961-1) [14]

Table 2.9 Specification of properties for wood chips

Master table			
Origin:	According to Table 2.8 Woody biomass (1)		
Traded Form	Wood chips		
Dimensions (mm)	Method: EN 15149-1, sieves according ISO3310-1		
	Minimum 75 wt% in main fraction, mm[a]	Fines fraction, wt-% (<3.15 mm)	Coarse fraction, wt%, max. length of particle, mm
P16A[c]	3.15 < P < 16 mm	<12%	<3% > 16 mm and all < 30 mm
P16B[c]	3.15 < P < 16 mm	<12%	<3% > 45 mm and all < 120 mm
P45A[c]	8 < P < 45 mm	<8%[b]	<6% > 63 mm and maximum 3.5% > 100 mm, all < 120 mm
P45B[c]	8 < P < 45 mm[b]	<8%[b]	<6% > 63 mm and maximum 3.5% > 100 mm, all < 350 mm
P63[c]	8 < P < 63 mm[b]	<6%[b]	<6% > 100 mm, all < 350 mm
P100[c]	16 < P < 100 mm[b]	<4%[b]	<6 % > 200 mm, all < 350 mm

Normative

[a] The numerical values (P-class) for dimension refer to the particle sizes (at least 75 wt%) passing through the mentioned round hole sieve size (EN 15149-1). The cross sectional area of the oversized particles shall be P16 < 1 cm², for P45 < 5 cm², for P63 < 10 cm² and P100 < 18 cm²

[b] For logging residue chips, which include thin particles like needles, leaves and branches, the main fraction for P45B is $3.15 \leq P \leq 45$ mm, for P63 is $3.15 \leq P \leq 63$ mm, and for P100 is $3.15 \leq P \leq 100$ mm and amount of fines (<3.15 mm) can be maximum 25 wt%

[c] Property classes P16A, P16B, and P45A are for non-industrial and property class P45B, P63, and P100 for industrial appliances. In industrial classes P45B, P63, and P100 the amount of fines can be stated from the following F04, F06, F08

Specification of properties for wood chips (*continued*)

Normative	**Moisture content** (wt% as received)	Method EN 14774
	M10	≤10%
	M15	≤15%
	M20	≤20%
	M25	≤25%
	M30	≤30%
	M35	≤35%
	M40	≤40%
	M45	≤45%
	M50	≤50%
	M55	≤55%
	M55+	>55% (maximum value to be stated)

Ash, A (wt% of dry basis)	Method: EN 14775
A0.5	≤0.5%
A0.7	≤0.7%
A1.0	≤1.0%
A1.5	≤1.5%
A2.0	≤2.0%
A3.0	≤3.0%
A5.0	≤5.0%
A7.0	≤7.0%
A10.0	≤10.0%
A10.0+	>10.0% (maximum value to be stated)

Normative/Informative

Nitrogen, N (wt% of dry basis) (Method: EN 15104)

N0.3	≤0.3%	*Normative:*
N0.5	≤0.5%	Chemically treated biomass (1.2.2; 1.3.2)
N1.0	≤1.0%	*Informative:*
N2.0	≤2.0%	All fuels that are not chemically treated
N3.0	≤3.0%	(see the exceptions above)
N3.0+	>3.0% (maximum value to be stated)	

Chlorine, Cl (wt% of dry basis) (Method: EN15289)

Cl0.02	≤0.02%	*Normative:*
Cl0.03	≤0.03%	Chemically treated biomass (1.2.2; 1.3.2)
Cl0.07	≤0.07%	*Informative:*
Cl0.10	≤0.10 %	All fuels that are not chemically treated
Cl0.10+	>0.10% (maximum value to be stated)	(see the exceptions above)

Informative

Net calorific value as received, Q (MJ/kg or kWh/kg) (Method: EN 14918)

Minimum value to be stated

Bulk density (BD) as received (kg/m^3) (Method: EN 15103)

BD150	>150
BD200	>200
BD250	>250
BD300	>300
BD350	>350
BD400	>400
BD450	>450
BD450+	>450 (minimum value to be stated)

Recommended to be stated if traded by volume basis

Ash melting behavior (°C) Method: EN 15370-1

Deformation temperature, DT should be stated

Note   Special attention should be paid to the ash melting behavior for some biomass fuels, for example eucalyptus, poplar, short rotation coppice.

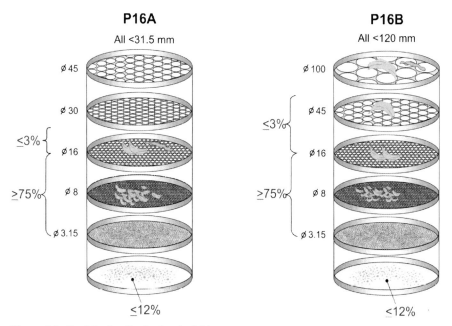

Figure 2.9 Particle size distribution for P16

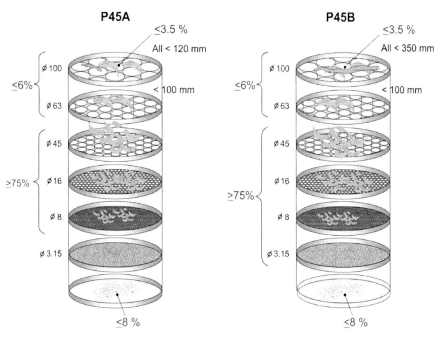

Figure 2.10 Particle size distribution for P45

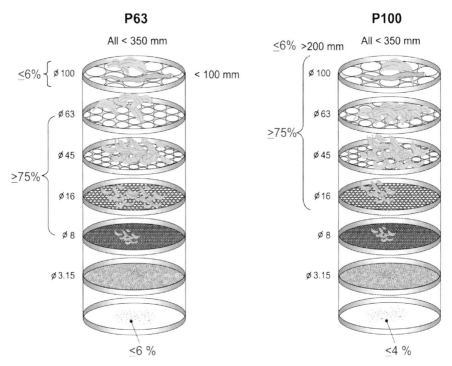

Figure 2.11   Particle size distribution for P63 and P100

References

1. Alakangas E (2009) European standards for solid biofuels – case wood pellets and wood chips, Riga 6 May 2009, Wood combustion and standards, Proc. Environ Climate Technol 13(2), pp 7–20
2. Alakangas E, Wiik C, Rathbauer J, Sulzbacher L, Kilgus D, Baumbach G, Grammelis P, Malliopoulou A, Naoum M, van Erp F, van Asselt B (2008) Used wood and chemically treated industrial wood residues and by-products in the EU. Part 2. Catalogue of used wood examples, BioNormII – Pre-normative research on solid biofuels for improved European standards, Project no. 038644, DIV6-Part 3. (www.bionorm2.eu)
3. Alakangas E (ed) (2009) Summary report of combustion test, BioNormII – Pre-normative research on solid biofuels for improved European standards, Project no. 038644, DIV7-Part 7. (www.bionorm2.eu)
4. Alakangas E (2009) Feedback on EN 14961 standards from industry and workshops BioNormII – Pre-normative research on solid biofuels for improved European standards, Project no. 038644, DIV6-Part 1. (www.bionorm2.eu)
5. Alakangas E, Wiik C, Lensu T (2007) CEN 335 – Solid biofuels, feedback from market actors, EUBIONET report – VTT Report VTT-R-00430-07, Jyväskylä 2007.
6. Alakangas E (2005) Experiences of using solid biofuel standards in biofuel trade and production, Proceedings. of the 14th European Biomass Conference, pp 17–21 October 2005, Paris,
7. Alakangas E, Levlin JE, Valtanen J (2005) Classification, specification and quality assurance for solid biofuels. Bioenergy in Wood Industry 2005 – International Bioenergy Conference and Exhibition – 12–15 September 2005, Jyväskylä, Finland, Proc. pp 307–312

8. Alakangas E, Levlin JE, Valtanen J (2006) CEN technical specifications for solid biofuels – fuel specification and classes, Biomass Bioenerg 30 11, pp 908–914
9. Alakangas E (2004) The European pellets standardisation – European Pellets Conference 3–4.3.2004, Wels Austria, Proceedings, pp 47–54
10. Alakangas E, Levlin JE, Valtanen J (2004) Fuel Specification and Classes, International Conference – Standardisation of Solid Biofuels, 6–7 October 2004, Leipzig, Germany, pp 57–66
11. EN 13556:2003 (2003) Round and sawn timber. Nomenclature of timbers used in Europe.
12. EN 303-5 (1999) , Heating boilers. Part 5: Heating boilers for solid fuels, hand and automatically stocked, nominal heat output of up to 300 kW. Terminology, requirements, testing and marking
13. EN15250 (2005), Slow heat release appliances fired by solid fuel – requirements and test methods
14. EN14961-1:2010 (2010) Solid biofuels – fuel specification and classes, Part 1 – General requirements. CEN (European Committee for Standardisation). January 2010.
15. Grammelis P, Malliopoulou A, Stamatis D, Lypiridis G (2009) Report of combustion tests, DIV.7 – Part 5.
16. Hartmann H, Turowski P, Ellner-Schuberth F, Winter S (2009) Fuel quality effects in wood log combustion – results from trials with a log wood boiler, TFZ, DIV.7 – Part 4,
17. EN 14588.2008 (2009) Solid biofuels – terminology, definitions and descriptions (final draft N68). January 2009.
18. EN 14961.2009 (2009) Solid biofuels – fuel specification and classes, Part 2 – Wood pellets for non-industrial use (draft document N192). May 2009.
19. EN 14961.2009 (2009) Solid biofuels – fuel specification and classes, Part 3 – Wood briquettetes for non-industrial use (draft document N194). May 2009.
20. EN 14961.2009 (2009) Solid biofuels – fuel specification and classes, Part 4 – Wood chips for non-industrial use (draft document N196). May 2009.
21. EN 14961.2009 (2009) Solid biofuels – fuel specification and classes, Part 5 – Wood logs for non-industrial use (draft document N198). May 2009.
22. EN 14961.2009 (2009) Solid biofuels – fuel specification and classes, Part 6 – Non-woody pellets for non-industrial use (draft document N200). May 2009.
23. EN 14961.2009 (2009) Solid biofuels – fuel quality assurance, Part 1 – General requirements (draft document N190). April 2009.
24. Oravainen H, Puolamäki K, Alakangas E (2009) Wood log combustion tests – slow heat release appliance. VTT. DIV.7-Part 1.
25. Oravainen H, Kolsi A, Alakangas E (2009) Wood log combustion tests – over-fire boiler, VTT, DIV.7-Part 2.
26. Sulzbacher L, Rathbauer J, Baumgartner H (2009) Pellet boiler combustion tests, FJ-BLT, DIV.7 – Part 3.
27. Wiik C, Alakangas E, Rathbauer J, Sulzbacher L, Kilgus D, Baumbach G, Grammelis P, Malliopoulou A, Naoum M, van Erp F, van Asselt B (2008) Used wood and chemically treated industrial wood residues and by products in the EU – Part 1, Classification, properties and practices, BioNormII – Pre-normative research on solid biofuels for improved European standards, Project no. 038644, DIV6-Part 2. (www.bionorm2.eu)

Chapter 3
Biomass-Coal Cofiring:
an Overview of Technical Issues

Larry Baxter

Abstract This investigation explores the reasons for and technical challenges associated with co-combustion of biomass and coal in boilers designed for coal (mainly pulverized coal) combustion. Biomass-coal co-combustion represents a near-term, low-risk, low-cost, sustainable, renewable energy option that promises reduction in effective CO_2 emissions, reduction in SO_x and often NO_x emissions and several societal benefits. Technical issues associated with cofiring include fuel supply, handling and storage challenges, potential increases in corrosion, decreases in overall efficiency, ash deposition issues, pollutant emissions, carbon burnout, impacts on ash marketing, impacts on SCR performance and overall economics. Each of these issues has been investigated and this presentation summarizes the state-of-the-art in each area, both in the US and abroad. The focus is on fireside issues. While each of the issues can be significant, the conclusion is that biomass residues represent possibly the best (cheapest and lowest risk) renewable energy option for many power producers.

3.1 Introduction

Cofiring biomass with coal simultaneously provides among the most effective means of reducing net CO2 emissions from coal-based power plants and among the most efficient and inexpensive uses of biomass. Recent reviews of cofiring experience identify over 100 successful field demonstrations in 16 countries that use essentially every major type of biomass (herbaceous, woody, animal-wastes and anthropomorphic wastes) combined with essentially every rank of coal and combusted in essentially every major type of boiler (tangential, wall, and cyclone

L. Baxter (✉)
Brigham Young University,
Provo, UT 84601, USA
e-mail: larry_baxter@byu.edu

fired) [1, 2]. Those countries that actively seek global climate change mitigation strategies rank cofiring as among the best (lowest risk and least expensive) options. Nevertheless, there remain substantial uncertainties associated with long-term implementation of this biomass technology. This document principally focuses on recent progress resolving technical issues associated with cofiring, with this introduction summarizing many of the primary motivations for pursuing this technology, including several that appear to be overlooked by current governmental and industrial strategic plans and policies.

The forest products industry and farming generate residues whose use as fuel represents among the most socially and environmentally beneficial biomass resources. Energy crops – crops harvested solely for their energy content – represent additional potential fuel resources in the total biomass energy potential but have increased environmental liability when one considers an entire life cycle [3]. So long as the farming and forest products industries that produce these residues and energy crops conduct themselves in a sustained manner, growing new plants at a rate greater than or equal to the harvest rate, there is no net increase in the atmospheric CO_2 associated with their use as fuel. Furthermore, residues not used as fuel generally decay to form CO_2 and often smaller quantities of other much more potent greenhouse gases. Therefore, redirecting the residues into a fuel stream in some cases decreases net greenhouse gas emissions even without counting the displacement of the fossil-derived fuels. However, the largest greenhouse gas contribution comes from the displacement of fossil fuels. The greenhouse gas reduction potential of biomass is directly associated with its sustainable production and is in no way dependent on that production being exclusively dedicated to power production. Indeed, sustainably produced residues exhibit greenhouse gas, environmental and economic benefits as fuels that generally exceed those of dedicated energy crops [3] – a point that appears to be overlooked by many governmental and industrial incentive programs. Energy crops represent economically and technically more challenging fuels than most residues but are also effective in reducing CO_2 when sustainably grown.

Addition of biomass to a coal-fired boiler does not impact or at worst slightly decreases the overall generation efficiency of a coal-fired power plant [4, 5]. Some of the more significant potential sources of efficiency decrease include use of non-preheated air in biomass burners/injectors, increased parasitic losses associated with generally more energy intensive fuel preparation and handling and increased moisture content in the fuel. The first issue would likely be eliminated in a permanent installation as contrasted with a short-term demonstration test. The remaining issues are highly fuel dependent and in any case would have less impact on efficiency calculated in the European tradition (lower heating value basis) than in the US and Australian tradition (higher heating value basis). In general, if all of the efficiency losses associated with biomass cofiring were allocated to only the biomass fraction of energy input, they would represent a 0–10% loss in biomass conversion efficiency compared to coal [6–13]. That is, biomass-coal cofiring results in biomass conversion efficiencies ranging from 30 to 38% (higher-heating value basis), easily exceeding efficiencies in dedicated biomass systems and rivalling or exceeding the estimated efficiencies of many future, advanced biomass-based systems. Therefore, commercialization of cofiring technologies offers among the best short-term and long-term solutions to greenhouse gas emissions reduction from power generation. Since cofiring is not an option in all localities, a robust biomass utilization strategy requires development of alternative technologies as well. However, the effectiveness of the cofiring option, combined with its low cost and low technical risk, should place it high on a priority list of institutions considering an array of greenhouse gas options.

Cofiring installation costs in many power plants are \$50–300/kW of biomass capacity [4, 5]. These low costs are achievable primarily because cofiring makes use of the existing infrastructure of a power plant with minimal infrastructural changes. These costs compare favourably with essentially any other available (hydropower being regarded as largely unavailable) renewable energy option. However, with rare exceptions, cofiring biomass will be more expensive than fossil energy. Cofiring usually displaces fossil power without increasing total capacity, so the capital costs with which to compare the previously quoted numbers is \$0/kW rather than the more typical \$900/kW for coal. In cases where additional capacity is anticipated, capital costs for cofiring are much higher when, for example, induced draft fans and other common capacity limiting subsystems must be replaced or upgraded. Operating costs are also typically higher for biomass than for coal. The most sensitive factor is the cost of fuel, resulting in energy crops suffering large economic disadvantages relative to residues. Even if the fuel is nominally free at the point of its generation (as many residues are), its transportation, preparation and on-site handling typically increase its effective cost per unit energy such that it rivals and sometimes exceeds that of coal. A general conclusion is that biomass cofiring is commonly slightly more expensive than dedicated coal systems. If there is no motivation to reduce CO_2 emissions, the rationale for cofiring is difficult to establish. However, the biomass component of cofiring represents renewable, essentially CO_2-neutral energy. In this respect, the more relevant cost comparison is that of cofiring with other renewable options. In this comparison, cofiring represents by far the cheapest means of renewable power generation in a large fraction of situations where it is feasible, feasibility being indicated by biomass resources and coal-based power plants available in the same region. Cofiring also represents a dispatchable, rapidly deployable, low-risk, regionally indigenous and inherently grid-compatible energy source, all significant advantages for overall grid management and power systems planning.

Cofiring represents a short-development-time, low-cost (compared to other renewable options), low-risk, high-social-benefit, energy option badly needed in energy markets of nearly every developed and many developing countries. The technology has been demonstrated at commercial scales in essentially every (tangentially fired, front-wall fired, back-wall fired, dual-wall fired and cyclone) boiler type, combined with every commercially significant (lignite, subbituminous coal, bituminous coal, and opportunity fuels such as petroleum coke) fuel type, and with every major category of biomass (herbaceous and woody fuel types generated as residues and energy crops). However, there are few long-term tests or fully commercialized preparation or handling systems.

The major technical challenges associated with biomass cofiring include:

1. fuel preparation, storage, and delivery;
2. ash deposition;
3. fuel conversion;
4. pollutant formation;
5. corrosion;
6. fly ash utilization;
7. impacts on SCR systems;
8. formation of striated flows.

Previous reports have focused on many of these issues (primarily the first five), the conclusions of which are summarized here. Three properties of biomass impact its preparation, storage and handling properties. Biomass has low bulk energy density, is generally moist and strongly hydrophilic and is non-friable. Biomass heating values are generally slightly over half that of coal, particle densities are about half that of coal and bulk densities are about one fifth that of coal. This results in an overall fuel density roughly one tenth that of coal. Consequently, cofiring biomass at a 10% heat input rate results in volumetric coal and biomass flow rates of comparable magnitudes. Consequently, biomass demands shipping, storage and on-site fuel handling technologies disproportionately high compared to its heat contribution.

Biomass produces a non-friable, fibrous material during comminution [6, 8–12]. It is generally unfeasible (and unnecessary) to reduce biomass to the same size or shape as coal. In many demonstration plants, biomass firing occurs with particles that pass through a 1/4-inch (6.4 mm) mesh, these measurements indicating a size distribution dominantly less than about 3 mm. Depending on the type of biomass and preparation technique, average aspect ratios of these particles range from three to seven, with many particles commonly having much higher aspect ratios. Such particles have very low packing densities and create challenges when pneumatically or otherwise transporting biomass fuels.

Ash deposit formation from biomass ranges widely, both in absolute terms and compared to coal [14–20]. In general, herbaceous materials potentially produce high deposition rates while many forms of wood waste produce relatively minor deposit rates. Treatment of the fuel by water leaching or other techniques impacts these results dramatically, consistent with the idea that alkali (potassium) and chlorine play a major role in the process. Deposit formation is a major consideration in fuel and boiler selection, but the majority of commercial systems in the US use wood-based materials that have relatively low deposition potential.

These large and non-spherical particles pose challenges for fuel conversion efficiency. Coal particles of such size would not nearly burnout in a coal boiler, but there are compensating properties of biomass. Biomass yields a much higher fraction of its mass through devolatilization than does coal [21, 22]. Typically biomass of the size and under the heating rates typical of pc-cofiring yields 90–95% of its dry, inorganic-free mass during devolatilization, compared with 55–60% for most coals. Devolatilization occurs rapidly and is temperature driven and therefore most biomass fuels will yield at least this fraction of mass so long as they are entrained in the flue gases. Biomass particles too large or dense to be entrained sometimes enter the bottom ash stream with little or no conversion beyond drying. However, these are generally the exception for well-tuned fuel preparation systems. Second, the low particle densities help biomass particles oxidize at rates much higher than coal. However, excessive moisture or excessive size particles still may pose fuel conversion problems for biomass cofiring despite these mitigating effects. Finally, the shape of biomass particles promotes more rapid combustion than the typically spherical shape of coal particles [23–25].

Pollutant formation and other gaseous emissions during biomass cofiring exhibits all the complexities as do the same issues for coal combustion [10, 26–28]. SO_x generally decreases in proportion to the sulphur in the fuel, which is low for many (but not all) biomass fuels. NO_x may increase, decrease or remain the same, depending on fuel, firing conditions, and operating conditions. However, the NO_x chemistry of biomass shows the same, complex but conceptually well understood behaviour as NO_x chemistry during coal combustion with the exception that biomass appears to produce much higher NH_3 content and a lower HCN content as a nitrogen-laden product gas compared to coal. Some of the commercially most mature biomass fuels, notably wood, contain relatively little fuel nitrogen and cofiring with such fuels tends to decrease total NO_x. However, general industrial experience is not consistent with the sometimes suggested truism [4] that NO_x reduction when cofiring biomass exceeds the fuel nitrogen displacement effect by 10%. Biomass fuels also commonly contain more moisture than coal, decreasing peak temperatures and leading to commensurate decreases in NO_x.

Results from the issues summarized above are illustrated in this document. However, the general conclusion does not change; Cofiring biomass with coal introduces several significant issues into boiler operation that have the potential for deleterious effects but none of these issues represents an insurmountable obstacle for biomass.

3.2 Fuel Characteristics

The biomass fuels considered here range from woody (ligneous) to grassy and straw-derived (herbaceous) materials and include both residues and energy crops. Woody residues are generally the most common fuels of choice for coal-fired boilers while energy crops and herbaceous residues represent future fuel resources and opportunity fuels, respectively. Biomass fuel properties differ significantly from those of coal and also show significantly greater variation. As examples (see Figures 3.1 and 3.2), ash contents vary from less than 1% to over 20% and fuel nitrogen varies from around 0.1% to over 1%. Other notable properties of biomass relative to coal are a generally high moisture content (usually greater than 25% and sometimes greater than 50% as-fired, although there are exceptions), potentially high chlorine content (ranging essentially from zero to 2.5%), relatively low heating value (typically, half that of hv bituminous coal) and low bulk density (as, as little as one tenth that of coal per unit heating value). These and other properties must be carefully considered for successful implementation of cofiring.

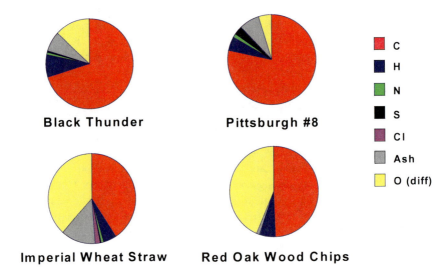

Figure 3.1   Typical ultimate analyses of biomass and coal

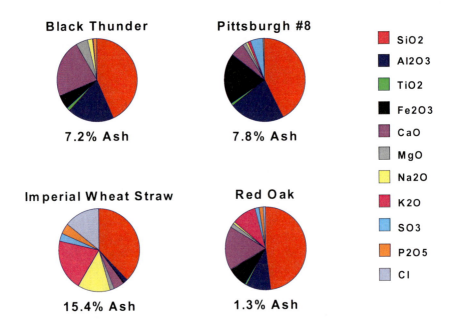

Figure 3.2   Typical variation in inorganic composition of biomass and coal fuels

3.3 Fuel Preparation and Transportation

Biomass is generally not as friable as coal and different equipment commonly comminutes biomass, generally at higher specific energy cost, compared to coal. The most commonly used comminution technologies include fuel hogs, tub grinders and screens of various types. It is theoretically possible to reduce biomass to arbitrarily low sizes by recycling the large material in a flow loop, but the cost of reducing biomass to sizes less than about 1/4 inch become exponentially higher.

Figure 3.3 illustrates one potential result of the substantially larger and different fuel properties of biomass relative to coal [29–31]. These data come from a pilot-scale, entrained-flow facility with a swirl-stabilized burner. The photographs indicate the appearance of pure biomass straw prepared for cofiring of a blend of the coal and straw, and of the particles sampled at the indicated distances from the burner in the biomass flame. As the flame progresses, clearly the coal particles and biomass particles that contain fine structure burn preferentially, leaving large pieces of biomass that ostensibly come from the straw knees. These knee particles represent a small fraction of the total flow, but they persist long enough in the flame to alter its behaviour in significant ways (Figure 3.4).

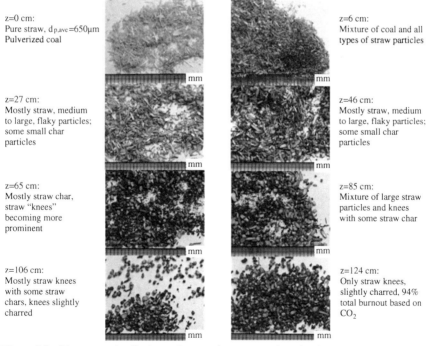

z=0 cm:
Pure straw, $d_{p,ave}=650\mu m$
Pulverized coal

z=6 cm:
Mixture of coal and all
types of straw particles

z=27 cm:
Mostly straw, medium
to large, flaky particles;
some small char
particles

z=46 cm:
Mostly straw, medium
to large, flaky particles;
some small char
particles

z=65 cm:
Mostly straw char,
straw "knees"
becoming more
prominent

z=85 cm:
Mixture of large straw
particles and knees
with some straw char

z=106 cm:
Mostly straw knees
with some straw
chars, knees slightly
charred

z=124 cm:
Only straw knees,
slightly charred, 94%
total burnout based on
CO_2

Figure 3.3 Biomass-coal cofired samples at various stages of combustion sampled from a pilot-scale, swirl-stabilized burner. The large and relatively dense biomass particles originating from the straw knees develop a secondary flame after the first flame supported by the coal and the remaining biomass fuel fraction. The knees are clearly evident in the last several particle samples [29]

Biomass fuels are hygroscopic, have bulk energy densities sometimes more than an order of magnitude lower than that of coal, and have shapes that lead to bridging and compaction in many fuel handling systems. For these reasons, biomass fuel preparation and handling are significantly more difficult than for an equivalent coal system and are generally best done in separate systems. The most common exceptions include biomass that has already been highly processed to an appropriate size (sander dust, some types of sawdust, torrefied fuels and some pellets) and very dense and relatively brittle biomass (some nut shells). When the fuels are mixed, specific care must be taken to prevent bridging and plugging in hoppers, around corners, *etc.* Generally, biomass is best prepared and handled as a separate fuel rather than being mixed with coal unless the biomass is already in a

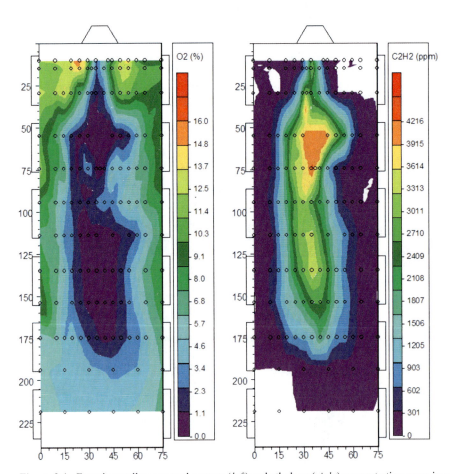

Figure 3.4 Experimentally measured oxygen (*left*) and ethylene (*right*) concentration maps in a down-fired, axisymmetric, pilot-scale combustor. The numbers along the combustor sides represent mm in distance. Combustion of straw knees creates an oxygen decrease and ethylene increase along the centre line starting at about 125 mm, as indicated from the particle samples displayed earlier [29, 32]

form compatible with the coal delivery system or in low enough concentration to represent a small perturbation of the coal system. However, the typical factor of 10 or more difference in the bulk energy density of biomass and coal means it requires relatively little biomass to present a significant perturbation on fuel handling systems.

Finally, biomass shape and size significantly alter particle combustion characteristics relative to those of coal particles [33–35]. Coal particles generally are small enough that they develop minimal internal temperature gradients. Biomass particles, on the other hand, are large and develop very large gradients. Figure 3.5 illustrates measurements and model predictions of particle internal and surface temperatures and mass losses. The data come from a special-build furnace operating, in this case, with gas and wall temperatures of 1050 K and 1273 K, respectively and an inert (N_2) environment. The initial particle diameter, moisture con-

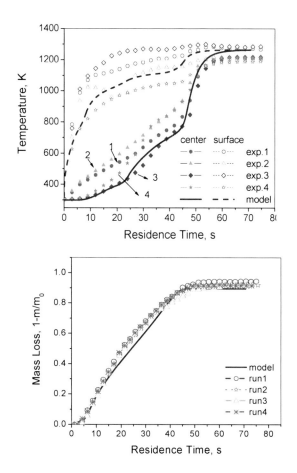

Figure 3.5 Experimental and measured data describing particle centre and surface temperature and mass loss as a function of time [33, 36–38]

tent and aspect ratio are 9.5 mm, 6% and 4, respectively. As shown, even under these mild heating conditions relative to a flame, the internal temperature gradient is very large during most of the particle history. The internal temperature data shown as 1 and 2 suffer from conduction along thermocouple leads whereas those of 3 and 4 were designed to minimize these effects. As seen, sophisticated model predictions capture the relatively complex temperature behaviour, being influenced strongly by moisture content, internal heat transfer, transpiration blowing effects on heat and mass transfer, *etc.* However, this behaviour differs markedly from that for a coal particle under similar conditions, and the temperature gradients and differences in behaviour become more severe at the much higher heat transfer rates experienced in flames.

Figure 3.6 illustrates the influence of more sophisticated predictions on particle behaviour. Two sets of model predictions, the first a sophisticated model and the second a model with coal-like assumptions (isothermal, spherical particle, *etc.*), appear with experimental data. The thermocouple wire shielding technique used to improve the early residence time data in Figure 3.5 could not be used in this experiment, so there exists some differences in the early between the model and the data at the centre, with the model being more reliable than the data (because of lead conduction). Otherwise, the sophisticated model predicts the data reasonably well at both the centre and the surface. The model with coal assumptions is in large error, even under these relatively mild heating conditions (1050 K and 1273 K gas and wall temperatures) and for this very dry biomass (6% moisture). Furthermore, this is a nearly spherical particle, and the shape also has a profound impact on reaction rate. With more rapid heating/higher gas and radiative temperatures and with more typical biomass moisture and shape, the differences would be

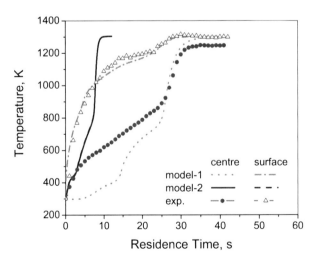

Figure 3.6 Experimental and measured data describing particle centre and surface temperature as a function of time compared to the predictions using model assumptions typical of a pulverized coal model (Model 2, *solid line*) [33, 36–38]

much larger still. The model with coal-type assumptions heats more rapidly primarily because the surface temperature remains lower in the model than it is in reality owing to the isothermal assumption. This comparison shows how important the shape, size and multidimensional effects are on particle behaviour and that these cannot be capture with traditional models used for coal particles. However, this result should not be misinterpreted to suggest that an actual coal particle heats up more rapidly than a biomass particle. Coal particles are much more dense that biomass and an actual coal particle of the same initial size (11 mm in this case) would heat up far more slowly, in large measure because it is more dense and because it also would experience the internal temperature gradients and other features that would slow its heating considerably compared to the isothermal assumption.

Biomass particles generate more complex temperature patterns than the radial temperature gradients measured and modelled here suggest. Figure 3.7 illustrates particle surface temperature measurements of a burning, nearly spherical, biomass particle using a recently innovated technique. The bottom right panel is a three-dimensional reconstruction of these data, also using a recently innovated technique. In this bottom-right panel, the bottom edge of the particle represents its leading edge. As seen, there is a large (approximately 500 K) temperature gradient from the hot spots near its leading edge to the cool spot in the wake of the particle. This gradient is along the surface, as opposed to from centre to surface as were the previous ones. These experimentally measured patterns suggest that combustion may not proceed relatively uniformly on all particle surfaces, as is generally assumed, but may proceed largely on the windward side of the particle, with the leeward side helping to vent the products of combustion.

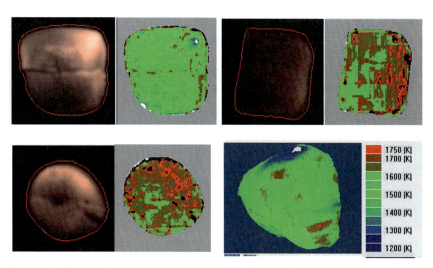

Figure 3.7 Three orthogonal images and corresponding pixel-by-pixel temperature maps of a burning biomass particle and a three-dimensional reconstruction of the particle shape and surface temperature [33, 36–38]

The guidelines for fuel preparation are to prepare the fuel using equipment designed specifically for biomass feed systems and to separate handling and transport lines for the biomass, except in circumstances of already highly processed fuels or very low loading. Large, aspherical biomass particles burn differently compared to typical coal particles, and these differences lead to differences in flame characteristics and particle properties that influence biomass fuel placement and burner behaviour.

3.4 Implementation

Biomass cofiring with coal usually represents among the least expensive and most efficient biomass to energy conversion options for renewable energy at coal-fired power plants (Figure 3.8). As mentioned above, biomass particles cannot realistically be reduced to the same size as coal particles. Therefore, they typically enter a boiler at much larger sizes. If they are injected at low burner levels, this leads to significant fractions falling into the bottom ash hopper without combusting. If they enter at the highest burner levels, they do not have sufficient residence time to combust completely prior to reaching the convection pass. Therefore, biomass injection commonly occurs in mid-level burners and, when possible, not in wall burners near the corners, avoiding biomass particle impaction on walls.

Despite the generally chaotic appearance of flow inside boilers, flows from different burners do not mix well. This is most clearly evident from the generally successful attempts at burner imbalance measurements based on oxygen concentration gradients near the bag house/precipitator entrance. This implies that flows in boilers are largely striated, meaning that although the overall biomass percentage in the boiler is low, there may be regions of the boiler that see nearly the same biomass percentage in the flue gas as occurs in an individual burner. Since there are beneficial synergistic interactions between coal and biomass flue gases, it is generally good practice to combine some coal with biomass in each burner. An example of the synergies this encourages is the sulphur mitigation of potential corrosion problems caused by some biomass. A reasonable strategy is to fire at least 50% coal in each burner that contains biomass.

Finally, biomass pneumatic transport is much more abrasive and erosive than coal transport. Generally, all pipe/duct corners in biomass lines require reinforcement with concrete to prevent erosive failure.

3.5 Pollutant Production

Pollutant production investigated here includes emissions of both SO_x and NO_x. SO_x emissions almost uniformly decrease when firing commercially significant types of biomass, often in proportion to the biomass thermal load. An additional incremental reduction beyond the amount anticipated on the basis of fuel sulphur content is sometimes observed and is based on sulphur retention by alkali and alkaline earth metals in the fuels. Some forms of biomass contain high levels of these materials. The SO_x emissions are relatively straightforward and are not illustrated in detail here.

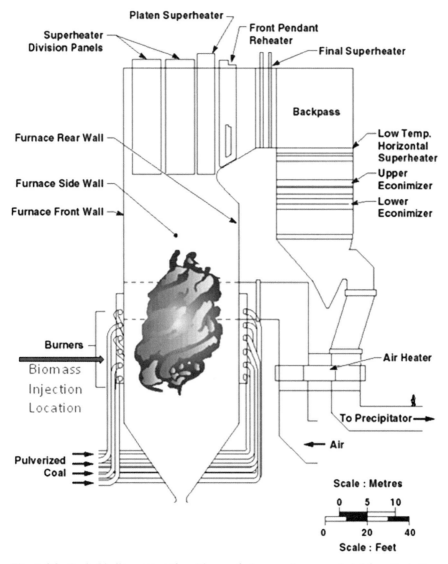

Figure 3.8 Typical boiler cross section. Biomass fuels generally are precluded from the bottom and top burner levels and frequently are co-injected with coal in mid-level burners, commonly away from the walls

NO$_x$ emissions are more difficult to anticipate. Experimental characterization of NO$_x$ emissions during combustion of neat coal, neat biomass, and various blends of the fuels combustion in a pilot-scale facility illustrate that NO$_x$ emissions from biomass can either exceed or be less than those of coal. Figure 3.9 illustrates data for an herbaceous fuel, showing that the ppm concentrations of NO$_x$ vary in a fairly complex manner with oxygen content. Low-nitrogen wood fuels typically

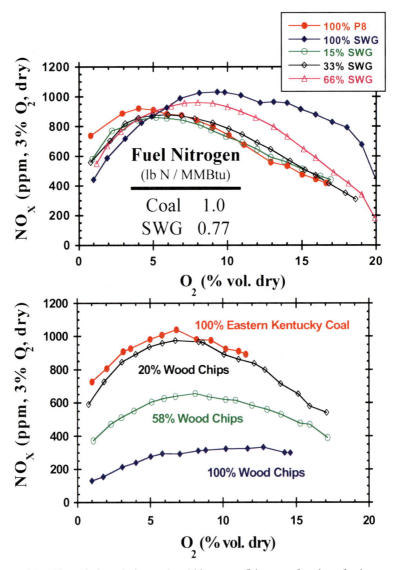

Figure 3.9 NO$_x$ emissions during coal and biomass cofiring as a function of exit gas oxygen content. These experiments do not use low-NO$_x$ burners, reburning, air/fuel staging or other NO$_x$ mitigation techniques [39]

produce much lower NO$_x$, most commonly uniformly less than that of coal as a function of oxygen content. When analyzed on a lb NO$_x$ per unit energy production basis instead of a ppm basis, NO$_x$ emissions from biomass fuels increase disproportionately compared to coal. Still, wood fuels generally produce lower NO$_x$ emissions than coal and herbaceous fuels may be higher or lower, depending on overall oxygen concentration and fuel nitrogen content.

The large difference in fuel oxygen contents between biomass and coal suggest that blends of coal and biomass could produce quite different results than would be expected based on the behaviour of the individual fuels. However, our data suggest there is no significant chemical interaction between the off-gases. Figure 3.10 illustrates a comparison of measured NO$_x$ emissions for a variety of coal-biomass blends with those predicted from the behaviour of the pure fuels. Points that fall along the diagonal indicate no significant interaction. Importantly, all of these experiments were conducted without low-NO$_x$ burners, fuel/air staging, reburning, *etc.* Generally, NO$_x$ emissions from blends of coal and biomass interpolate quite accurately between the measured behaviours of the neat coal and biomass fuels if no low-NO$_x$ burner, fuel staging, or boiler technology is used. Since biomass produces a significantly larger volatile yield than coal, there is potential for biomass to be effective in creating large fuel-rich regions useful for NO$_x$ control. The biomass fuels best suited for use in pc boilers are woods, most of which reduce total NO$_x$ emissions significantly below that from coal.

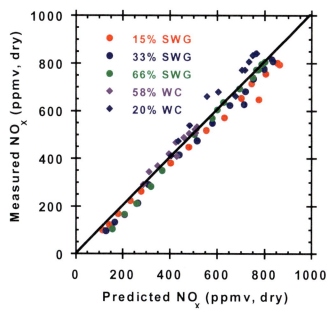

Figure 3.10 Comparison of measured NO$_x$ concentrations to concentrations interpolated (predicted) from the measured behaviour of the pure fuels. Data along the *diagonal line* indicate no interaction between the fuels [39]

The data reported thus far use highly idealized flows lacking the complexities of swirl, recirculation zones and strong stoichiometric gradients that exist in real systems. When these complexities are added, NO_x behaviour becomes more complex. For example, Figure 3.11 summarizes data for a variety of fuels under simi-

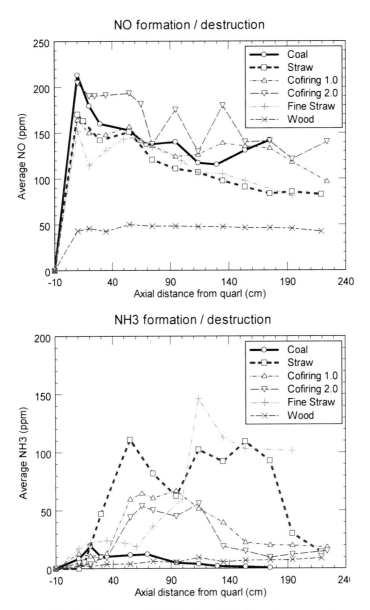

Figure 3.11 Radially averaged NO (*top*) and NH_3 (*bottom*) concentrations in an axisymmetric, swirl-stabilized, down-fired combustor [29]

lar combustion conditions, indicating that biomass NO chemistry differs from coal in complex flows, mainly because of fluid mechanic and thermodynamic effects [29–32]. In this case, the difference in the final flame NO_x concentrations has to do with NH_3 concentrations and their mixing with the flue gas. Biomass fuel nitrogen primarily forms NH_3 whereas coal fuel nitrogen primarily forms HCN. The initial NO_x profiles of all the fuels except wood (which contains virtually no fuel nitrogen) are similar. The rise and decline is typical of low-NO_x burner designs in which a fuel-rich region helps reduce NO to N_2. Towards the end of the test section, where temperatures had dropped and gases had more thoroughly mixed, the ammonia in the straw-based fuels helped further reduce NO_x through processes in ways similar to SNCR combined with reburning – sometimes called enhanced reburning. This further reduced NO_x concentrations by about 30%. There is nothing more unusual or fundamentally different in these data than in traditional NO_x data except that some forms of biomass produce high NH_3 contents (in contrast to coal). The chemistry is the same but the outcomes depend in complex ways on the interplay between stoichiometry, temperature, and gas composition, as is generally the case with NO_x formation.

The guidelines derivable from this work relative to NO_x emissions include: (1) there is insignificant chemical interaction between the off gases from biomass and coal that would alter NO_x emissions; (2) NO_x emissions from the most well-suited biomass fuels for cofiring (wood residues) generally are lower than those from coal, leading to some overall NO_x reduction relative to coal during cofiring; and (3) the large volatile yield from biomass can be used to advantage to lower NO_x emissions during cofiring through well-established, stoichiometric-driven means.

3.6 Carbon Conversion

It is impractical to reduce most biomass fuels to the size of pulverized coal. A small fraction of such fuels, such as sander dust, is available in small sizes because of upstream processing. The great majority of fuels will require size reduction. Size reduction of biomass is nearly always more energy intensive than for coal. A concern regarding overall burnout of the biomass fuel arises as the sizes of pulverized coal particles are compared with those of practically achievable sizes for biomass fuels. Biomass and coal are consumed by both thermal decomposition (devolatilization) reactions and by char oxidation (Figure 3.6). A larger fraction of biomass is released as volatile gases during combustion (85–95% of initial particle mass) than is released from coal (50–65%). This large volatile yield occurs over a relatively short time (Figure 3.12) and significantly decreases the time required for complete combustion compared to a coal particle of similar size. The largest fraction of biomass' and coal's combustion history involves char oxidation. Experimental data indicate that biomass chars burn under strongly diffusion controlled conditions, as is consistent with theory. However, the rates of combustion differ from that of coal owing to its generally aspherical shape and lower char density, both of which effects can be reasonably well modelled (Figures 3.7 and 3.13). Furthermore, the slip velocity between char particles and local gas is higher for biomass than coal, increasing the effective residence time of a char particle for combustion. Moisture content also significantly impacts biomass burnout time. Devolatilization, while slower for biomass than coal, is generally much shorter than either drying or char combustion. The increased time for biomass devolatilization relative to coal is a consequence of the already discussed intra-particle temperature gradients in the relatively large biomass particles.

The guidelines relative to carbon conversion derivable from this work include: (1) particles prepared with top sizes greater than 3 mm (1/8 inch) will experience increasing difficulty completing combustion, with significant residual carbon expected at sizes greater than 6 mm (1/4 inch) as measured by the smallest dimen-

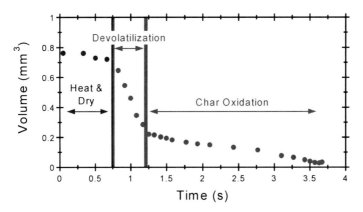

Figure 3.12 Combustion history of a typical biomass fuel (switchgrass in this case) illustrating the major stages of combustion. Coal has a similar history [39]

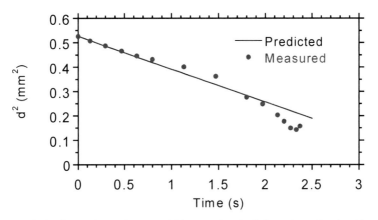

Figure 3.13 Comparison of measured biomass char (switchgrass) combustion history and that predicted by a recently developed theoretical model [39]

sion in the typically non-equant particles; (2) fuels with moisture contents exceeding 40% will need to be reduced further in size to achieve complete combustion; and (3) biomass char burning rates are controlled by geometry and size, not kinetics, making burning rates essentially fuel independent if size, shape, density and moisture contents are the same.

3.7 Ash Deposition

Ash deposit formation represents arguably the single most important combustion property impacting boiler design and operation for ash-forming fuels. Ash deposition rates from biomass fuels can greatly exceed or be considerably less than those of coal. Figure 3.14 illustrates rates of deposit accumulation during standardized experiments on simulated superheater tubes.

Absolute deposition rates from some herbaceous fuels exceed that of coal under identical conditions by about an order of magnitude whereas deposition rates for high-quality woods are nearly an order of magnitude less than that of coal. These trends are in part attributable to the ash contents of the fuels. When normalized for ash content differences, ash deposition efficiencies of herbaceous materials still exceed those of coal whereas those of wood are lower. These trends can be described in terms of ash particle sizes and chemistry. Deposition rates from blends of coal and biomass lie between the observed rates for the neat fuels but are gener-

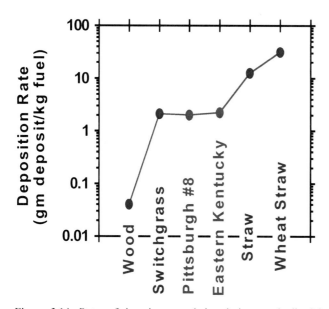

Figure 3.14 Rates of deposit accumulation during standardized investigations on simulated superheater tubes in the Multifuel Combustor [39]

ally lower than one would expect if interpolating between the behaviours of the neat fuels. Experimental evidence supports the hypothesis that this reduction in ash deposition occurs primarily because of interactions between alkali (mainly potassium) from the biomass and sulphur from the coal. Some of these data are presented in the discussion of corrosion.

The guidelines relative to ash deposition include: (1) deposition rates should decline when cofiring wood or similar low-ash, low-alkali, low-chlorine fuels; (2) deposition rates should increase when cofiring high-chlorine, high-alkali, high-ash fuels, such as many herbaceous materials; and (3) deposition rates depend strongly on both individual fuel properties and interactions between the cofired fuels.

3.8 Corrosion

Figures 3.15 and 3.16 illustrate interactions between sulphur from coal and chlorine from biomass to mitigate corrosion in [19, 40, 41]. The principal result is that alkali chlorides that sometimes condense from chlorine-laden biomass fuel flue gases react with SO_2, generated primarily from coal, to form alkali sulphates, which are significantly less corrosive. Figure 3.17 illustrates theoretical (equilibrium) predictions indicating that this only occurs under oxidizing conditions. Under reducing conditions, chlorides, not sulphates, are the stable form of alkali species under typical boiler heat transfer conditions. Therefore, the ameliorating effects of coal-derived sulphur on corrosion during cofiring do not occur in regions of boilers where deposits are exposed to reducing conditions. Further experimental data indicate that, even under oxidizing conditions, chlorine deposits may persist for many hours if deposit temperatures are very cool, reducing the kinetic rates of conversion to sulphates [19, 42–52].

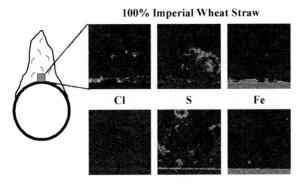

Figure 3.15 SEM images illustrating formation of chlorine layers on simulated boiler tubes and the effect of coal-derived sulphur during cofiring in eliminating the chlorine layers [19, 40, 41]

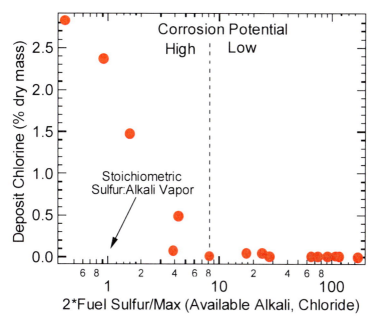

Figure 3.16 Results from systematic variation of fuel chlorine to sulphur ratios and the resulting chlorine content of deposits under standardized testing conditions [19, 40, 41]

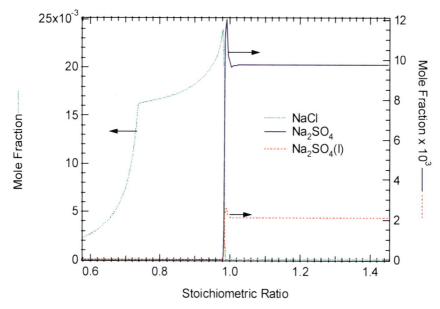

Figure 3.17 Illustration of predicted stoichiometric dependence of chlorine concentration in deposits

3.9 Fly Ash Utilization

The concrete market presents among the best fly ash utilization opportunities for coal-derived fly ash. However, the ASTM specification (ASTM Standard C618) for use of fly ash in concrete requires that the fly ash be derived entirely from coal combustion. Many processes in modern boilers result in coal fly ash mingled with other materials, including ammonia from pollutant control devices, sorbents or other injected materials from scrubbers, residual sulphur or other compounds from precipitator flue gas treatments, and fly ash from cofired fuels such as biomass. There is a broad, but not universal, recognition that the standard should be modified, but it is not clear what modifications should be made. Here some preliminary results regarding the impact of biomass-derived ash on concrete properties are presented [53].

This systematic investigation of the impact of biomass- and coal-derived fly ash on concrete involves both Class C (subbituminous) and Class F (bituminous) fly ash as well as similar fly ashes mingled with herbaceous and woody biomass fly ash. In all cases, 25% of the cement originally used in the concrete is displaced by fly ash, with the fly ash containing 0–40% biomass-derived material. Tests of concrete air entrainment, flexural strength, compressive strength, set time, freeze thaw behaviour and chlorine permeability determine the extent of the biomass impact. Only selected results are presented here and, as the tests require up to a year to conduct, all results are preliminary. The focus is on the herbaceous biomasses, since many woody fuels contain so little ash that practical cofiring is not likely to have a measurable impact on fly ash properties.

Figure 3.18 illustrates the impact of fly ash on the required amount of aerating agent to establish ASTM-compliant air entrainment levels in concrete [54–58]. Air

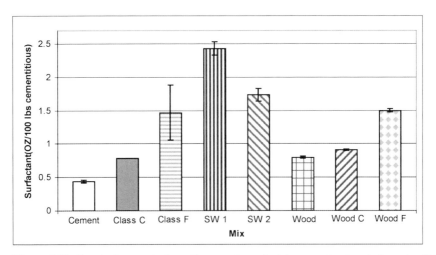

Figure 3.18 Required amount of aerating agent required to generate air entrainment within ASTM specifications for a variety of fly ash compositions [55–58]

entrainment in concrete is essential to prevent failure during freeze-thaw cycles. The amount of aerating agent increases with increasing herbaceous biomass content (the SW samples include 25% switchgrass with coal). This dependence arises from the effect of water soluble components (higher in herbaceous biomass than in coal fly ash) tying up the aerating agent (generally surfactants), preventing them from forming films that support bubble growth. The impact illustrated is of minor economic concern but is of process concern. That is, if fly ashes from cofired units were treated the same way as fly ashes from coal, the resulting concrete would likely fail under freeze thaw cycles. Increasing the surfactant to an acceptable level is of little economic impact, but failure to recognize the need to adjust it is of major impact.

Figure 3.19 illustrates the impact of biomass-coal commingled fly ash on flexural strength. In these test little significant difference is seen among the various samples with the possible exception of the pure wood sample (labelled Wood). The biomass-coal comingled ashes (all but the first three samples) perform similarly to pure coal fly ash (Class C and Class F) and to pure cement after 56 days. Additional data on set time, compressive strength, chlorine permeability, freeze thaw and many other similar tests indicate that all fly ashes delay set time by 2–4 h compared to concrete made from cement only but the biomass-containing fly ash does not delay set times significantly more than the non-biomass containing fly ash. Early compressive strength (in the first month or so) is compromised by all fly ashes, again with the biomass-containing fly ash similar to coal fly ashes. However, late strength (longer than 2 months or so) is enhanced by the presence of all fly ashes. Otherwise, there are no significant differences among the concrete samples made with comingled coal and biomass fly ash. The pure biomass fly ash samples, however, did not perform as well as the coal-biomass comingled ashes or the pure coal ashes.

In several important cases, the biomass fly ash actually outperforms coal fly ash. Figure 3.20 illustrates the performance of pure cement (top line) biomass-containing fly ashes (second group), pure coal fly ash (third group) and pure bio-

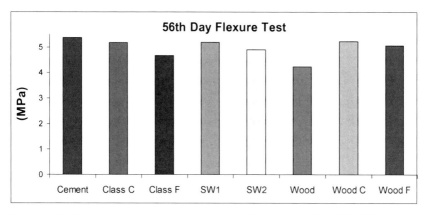

Figure 3.19 Flexural strength and its dependence on fly ash composition [55–58]

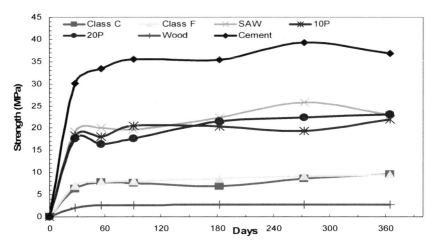

Figure 3.20 Pozzolanic reaction rates for a variety of samples [55–58]

mass ash (bottom line) under conditions where only the pozzolanic reaction (CaOH with fly ash) and not the cementitious reaction (calcium silicate hydration) is important. As seen, biomass enhances the pozzolanic reaction significantly, presumably because it contains amorphous rather than crystalline silica, the former being much more rapid to react with CaOH in the mix. This is the reaction that ultimately builds strength in concrete beyond that developed by the cementitious reaction. In these investigations, a CaOH-containing mixture reacted with the indicated material without large or small aggregate for up to a year as strength was monitored. Except in the pure cement case, there was no opportunity for cementitious reactions to occur. In another important series of experiments, the aggregate-silica reaction that leads to expansion and potential failure of concrete in the long term was similarly suppressed by biomass-containing fly ash as with pure coal fly ash. The biomass-containing sample performance was between that of Class F fly ash (best performing) and Class C fly ash (worst performing).

In conclusion, there appear to be only manageable impacts of biomass-containing fly ash on concrete properties based on these preliminary data, with the amount of aerating agent being an example of one issue that requires monitoring. Otherwise, biomass-containing fly ash behaves qualitatively similar to coal fly ash with no biomass in terms of structural and performance properties when incorporated into concrete.

3.10 Formation of Striated Flows

Many boilers do not mix flue gases effectively in furnace sections, resulting in gas compositions near the boiler exit that reflect burner-to-burner variations in stoichiometry and other properties. The impact of such behaviour during cofiring

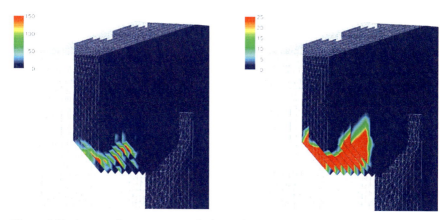

Figure 3.21 Impact of temperature, velocity and gas composition striations on two major classes of deposit formation mechanisms: impaction mechanisms (*left*) and boundary-layer mechanisms (*right*) [59]

can be an issue if one is hoping, for example, that sulphur from coal will mix with biomass-derived flue gases to ameliorate corrosion. Biomass is commonly injected in only a few burners. If the gases do not mix thoroughly, many regions of the boiler will be exposed to much higher biomass cofiring percentages than suggested by the overall average.

Advanced computational fluid mechanics models illustrate the impact of striations on ash deposition. Figure 3.21 illustrates the deposition patterns predicted on superheater tubes under conditions where such striations exist. As illustrated, there are large local variations in the rate of deposit accumulation. These arise from lack of complete mixing and striation in gas composition, gas temperature, gas velocity, particle loading and other similar properties (not illustrated). Such results are highly system dependent but are believed to be a common feature of biomass-coal combustion as well as both dedicated coal and dedicated biomass combustors.

3.11 Impacts on SCR Systems

Essentially, none of the cofiring demonstrations conducted in the US was performed on SCR-equipped boilers, but several tests from Europe have been conducted on such boilers. Some evidence from these tests is that cofiring biomass with coal results could deactivate SCR catalysts. The reasons for this deactivation are not definitive, but laboratory analyses confirm that alkali and alkaline earth metals are significant poisons to vanadium-based catalysts (which would include all commercial SCR systems) when the metals are in intimate association with the catalyst. Essentially, all biomass fuels contain high amounts of either alkali or alkaline earth metals or both as a percentage of ash. Some biomass fuels, however, have remarkably low ash contents, clean heartwood such as sawdust being a classical

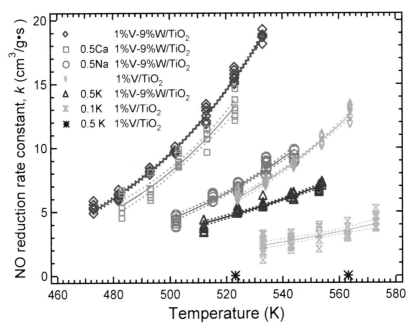

Figure 3.22 Laboratory measurements of SCR catalyst poisoning by materials contained in biomass [61–64]

example. It is possible that the commercially observed SCR deactivation arises from such poisoning or from catalyst fouling, which is also associated with such poisoning [13, 45, 60].

Laboratory and slip-stream commercial investigations add considerable information to this issue [61, 62]. Laboratory experiments clearly indicate the potential for SCR poisoning, as indicated in Figure 3.22 for a series of tests with varying amounts of material that occur in biomass in mobile forms and for varying catalyst compositions. The data show that alkali metals in particular (K and Na) but also alkaline earth metals (Ca) – in fundamental terms, anything that neutralizes Brønsted acid sites – can decrease reactivity significantly, up to the limit of completely removing catalyst activity. Catalysts that contain some tungsten (as almost all commercial catalysts do) are less susceptible but not immune to this deactivation. These data are collected under conditions where only catalyst poisoning is considered and poison impregnation in the catalyst was complete. Commercial system behaviour is more complex.

Figure 3.23 illustrates results from surface composition analyses from a catalyst exposed to the slipstream of utility combustor cofiring biomass that included alkali- and alkaline-earth-rich fuels. These normalized compositions show that this catalyst, which experienced significant deactivation, is enriched in sulphur, calcium and silica after exposure relative to the pre-exposure concentrations. However, the enrichment is limited to the outer surface and arises from an accumulation of calcium sulphate and silica from the fly ash. These results are consistent

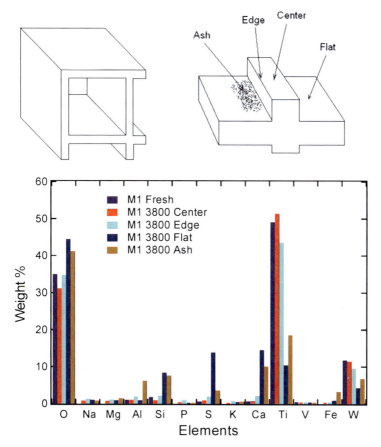

Figure 3.23 Concentrations of key elements (reported as oxides) in pre- and post-exposed catalyst from a slip-stream reactor [61–65]

with the general hypothesis that alkali and alkaline earth metals can cause catalyst deactivation, as shown above, but that surface fouling dominates the deactivation mechanism to a much greater extent than alkali poisoning and that alkali and alkaline earth materials do not appreciably impregnate the interior of most catalysts in practice.

The conclusion of these investigations is that SCR deactivation may be more of a concern with biomass-cofired systems, but that the deactivation mechanisms is likely more closely related to surface fouling on the catalyst, and possibly pore plugging, than to chemical poisoning. In many cases, such as clean wood fired with coal, the perturbation on ash chemistry associated with cofiring is small compared to the natural fluctuations in coal ash content. However, when cofiring large amounts of especially herbaceous biomass, or even large amounts of biomass in a single burner, all of some portion of the catalyst may see significant changes in composition and hence more fouling and possibly poisoning tendency.

This issue will become increasingly important as increased numbers of boilers install SCR systems to comply with lower NO_x emission limits. The authors are engaged (with others) in several investigations to explore more fully this phenomenon, including advanced laboratory and field tests.

3.12 Conclusions

Biomass cofiring with coal represents an attractive option for reducing greenhouse gas emissions from coal-fired boilers. In general, there are compelling reasons to pursue this option as reviewed in the Introduction. However, there are many issues that, if not carefully managed, could compromise the boiler or downstream processes. Results to date indicate that these are all manageable but that they require careful consideration of fuels, boiler operating conditions and boiler design.

References

1. Baxter L (2004) Biomass Cofiring Overview. 2nd World Conference and Exhibition on Biomass for Energy, Industry and Climate Protection
2. Koppejan J (2004) Introduction and overview of technologies applied worldwide. 2nd World Conference and Exhibition on Biomass for Energy, Industry and Climate Protection
3. Ouwens CD, Schonewille W, Kupers G (2002) Large-scale production of biomass derived Fischer–Tropsch liquids in the Rotterdam Harbor area – a case study. Proceedings of the Conference on the future for pyrolysis and gasification of biomass and waste
4. Baxter LL, Rumminger M, Lind T, Tillman D, Hughes E (2000) Cofiring Biomass in Coal Boilers: Pilot- and Utility-scale Experiences. Biomass for Energy and Industry: 1st World Conference and Technology Exhibition
5. Tillman DA, Hughes E, Gold BA (1994) Cofiring of biofuels in coal fired boilers: Results of case study analysis. 1st Biomass Conference of the Americas
6. Battista JJ, Hughes EE, Tillman DA (2000) Biomass cofiring at Seward Station. Biomass Bioenerg 19:419–427
7. Boylan DM (1996) Southern Company tests of wood/coal cofiring in pulverized coal units. Biomass Bioenerg 10:139–147
8. Hunt EF, Prinzing DE, Battista JJ, Hughes E (1997) The Shawville coal/biomass cofiring test: A coal/power industry cooperative test of direct fossil-fuel CO_2 mitigation. Energ Convers Manage 38:S551–S556
9. Junker H, Baxter LL, Robinson AL, Widell KE (1997) Cofiring Biomass and Coal: Plant Comparisons and Experimental Investigation of Deposit Formation. Engineering Foundation Conference on the Impact of Mineral Impurities on Solid Fuel Combustion
10. Surmen Y, Demirbas A (2003) Cofiring of biomass and lignite blends: resource facilities; technological and environmental issues. Energ Source 25:175–187
11. Swanekamp R (1995) Biomass co-firing technology debuts in recent test burn. Power 139:51–53
12. Tillman DA (2000) Biomass cofiring: the technology, the experience, the combustion consequences. Biomass Bioenerg 19:365–384
13. Wieck-Hansen K, Overgaard P, Larsen OH (2000) Cofiring coal and straw in a 150 MWe power boiler experiences. Biomass Bioenerg 19:395–409

14. Bakker RR, Jenkins BM, Williams RB, Carlson W, Duffy J, Baxter LL, Tiangco V (1997) Boiler performance and furnace deposition during a full-scale test with leached biomass. 3rd Biomass Conference of the Americas

15. Baxter LL (1993) Ash deposition during biomass and coal combustion: a mechanistic approach. Biomass Bioenerg 4:85–102

16. Nielsen HP, Baxter LL, Sclippab G, Morey C, Frandsen FJ, Dam-Johansen K (2000) Deposition of potassium salts on heat transfer surfaces in straw-fired boilers: a pilot-scale study. Fuel Elsevier Science Ltd Exeter England

17. Robinson AL, Buckley SG, Baxter LL (2001) Thermal conductivity of ash deposits 1: measurement technique. Energ Fuel 15:66–74

18. Robinson AL, Buckley SG, Yang NYC, Baxter LL (2001) Thermal conductivity of ash deposits 2: effects of sintering. Energ Fuel 15:75–84

19. Robinson AL, Junker H, Baxter LL (2002) Pilot-scale investigation of the influence of coal-biomass cofiring on ash deposition. Energ Fuel 16:343–355

20. Baxter LL, Richards GH, Ottesen DK, Harb JN (1993) In-situ, real-time characterization of coal ash deposits using Fourier-Transform Infrared-Emission Spectroscopy. Energ Fuel 7:755–760

21. Robinson AL, Baxter LL, Dayton D, Freeman M, Goldberg P (1997) Biomass-Coal Cofiring Experience Among Three Laboratories, NYSEG Review Meeting

22. Robinson A, Baxter LL, Freeman M, James R, Dayton D (1998) Issues Associated with Coal-Biomass Cofiring, Bioenergy '98

23. Sjostrom K, Chen G, Yu Q, Brage C, Rosen C (1999) Promoted reactivity of char in co-gasification of biomass and coal: synergies in the thermochemical process. Fuel 78: 1189–1194

24. Gera D, Mathur MP, Freeman MC, Robinson A (2002) Effect of large aspect ratio of biomass particles on carbon burnout in a utility boiler. Energ Fuel 16:1523–1532

25. Lu H, Scott J, Echols K, Foster P, Ripa B, Farr R, Baxter LL (2004) Effects of particle shape and size on black liquor and biomass reactivity. Science in Thermal and Chemical Biomass Conversion

26. Wu C, Damstedt B, Burt S, Tree D, Baxter L (2004) Fuel-nitrogen chemistry during combustion of low-grade fuels in a low-NO_x burner. Science in Thermal and Chemical Biomass Conversion

27. Robinson AL, Junker H, Baxter LL (1997) Pollutant formation, ash deposition, and fly ash properties when cofiring biomass and coal. Engineering Foundation Conference on the Economic and Environmental Aspects of Coal Utilization

28. Dayton DC, Belle-Oudry DA (1997) Bench-Scale Biomass/Coal Cofiring Studies. Engineering Foundation Conference on the Impact of Mineral Impurities on Solid Fuel Combustion

29. Damstedt B, Pederson JM, Hansen D, Knighton T, Jones J, Christensen C, Baxter L, Tree D (2007) Biomass cofiring impacts on flame structure and emissions. Proc Combust Inst 31:2813–2820

30. Wu C, Damstedt BD, Burt S, Tree DR, Baxter LL (2004) Fuel-nitrogen chemistry during combustion of low-grade fuels in a low-NO_x burner. Science in Thermal and Chemical Biomass Conversion

31. Wu C, Damstedt BD, Burt S, Tree DR, Baxter LL (2006) The study of fuel-nitrogen chemistry during combustion of low-grade fuels in a low-NO_x burner. Science in Thermal and Chemical Biomass Conversion, Victoria, British Colombia, Canada, pp 55–69

32. Damstedt BD, Hansen DC, Jones JJ, Christensen C, Johnson C, Jones T, Muff MV, Tree DR, Baxter LL (2007) Fuel-nitrogen chemistry in coal, biomass, and cofired flames. 5th US Combustion Meeting

33. Lu H, Ip E, Scott J, Foster P, Vickers M, Baxter LL (2010) Effects of particle shape and size on devolatilization of biomass particle. Fuel 89(5):1156–1168

34. Lu H, Ip LT, Mackrory A, Werrett L, Scott J, Tree D, Baxter L (2009) Particle surface temperature measurements with multicolor band pyrometry. AICHE J 55:243–255

35. Lu H, Robert W, Peirce G, Ripa B, Baxter LL (2008) Comprehensive study of Biomass particle combustion. Energ Fuel 22:2826–2839
36 Baxter LL, Ip LT, Lu H, Tree DR (2005) Distinguishing biomass combustion characteristics and their implications for sustainable energy. In Dally B (ed) 5th Asia-Pacific Conference on Combustion
37. Lu H, Roberts W, Werret L, Peirce G, Baxter LL (2006) Comprehensive study of biomass particle combustion. In: Raimo A (ed) 7th International Colloquium on Black Liquor Combustion and Gasification
38. Roberts WB, Lu H, Baxter L (2006) Effects of modeling assumptions on black liquor droplet combustion. In: Raimo A (ed) 7th International Colloquium on Black Liquor Combustion and Gasification
39. Robinson AL, Junker H, Buckley SG, Sclippa G, Baxter LL (1998) Interactions between coal and biomass when cofiring. Symposium (International) on Combustion Proceedings of the 1998 27th International Symposium on Combustion, Aug 2–7 1998 1:1351–1359
40. Junker H, Fogh F, Baxter L, Robinson A (1998) Co-firing Biomass and Coal: Experimental Investigations of Deposit Formation. 10th European Conference and Technology Exhibition
41. Robinson A, Junker H, Buckley SG, Sclippa G, Baxter LL (1998) Interaction between coal and biomass when cofiring. 27th Symposium (International) on Combustion/The Combustion Institute 1351–1359
42. Nielsen HP, Frandsen FJ, Dam-Johansen K, Baxter LL (2000) Implications of chlorine-associated corrosion on the operation of biomass-fired boilers. Progress in Energy and Combustion Science. Elsevier Science Ltd, Exeter, England
43. Robinson AL, Junker H, Buckley SG, Sclippa G, Baxter LL (1998) Interactions between coal and biomass when cofiring. Symposium (International) on Combustion Proceedings of the 1998 27th International Symposium on Combustion, Aug 2–7 1998 Combustion Inst, Pittsburg, PA, USA
44. Lokare S, Dunaway JD, Rogers D, Anderson M, Baxter L, Tree D (2004) Effects of fuel ash composition on corrosion deposits. In Proceedings of Science in Thermal and Chemical Biomass Conversion
45. Baxter L (2004) Biomass–coal Co-combustion: Renewable energy options for both industry and environmental groups. Science in Thermal and Chemical Biomass Conversion
46. Blander M, Pelton AD (1997) The inorganic-chemistry of the combustion of wheat–straw. Biomass Bioenerg 12:295–298
47. Dunaway D, Lokare S, Anderson M, Baxter L, Tree D (2006) Ash deposition rates for a suite of biomass fuels and fuel blends. Science in Thermal and Chemical Biomass Conversion
48. Dunaway JD, Lokare S, Rogers D, Moulton D, Junker H, Tree D, Baxter L (2002) Quantitatively Measured Ash Deposition Rates for a Suite of Biomass Fuels. Spring Meeting, Western States Section, The Combustion Institute
49. Junker H, Sander B, Stitt S, Tree D, Lokare S, Baxter L, Dunaway JD, Moulton D, Rogers D (2002) Agricultural Residues for Power Production. 12th European Biomass Conference, 2002.
50. Lokare S, Dunaway JD, Rogers D, Anderson M, Baxter L, Tree D (2006) Effects of Fuel Ash Composition on Corrosion Deposits. Science in Thermal and Chemical Biomass Conversion
51. Lokare S, Moulton D, Junker H, Tree D, Baxter L (2002) Effects of Fuel Ash Composition on Corrosion. Spring Meeting, Western States Section, The Combustion Institute
52. Lokare SS, Dunaway JD, Moulton D, Rogers D, Tree DR, Baxter LL (2006) Investigation of ash deposition rates for a suite of biomass fuels and fuel blends. Energ Fuel 20:1008–1014
53. Wang S, Fonseca F, Miller A, Llamazos E, Baxter L (2004) Biomass Fly Ash in Concrete. In proceedings of Science in Thermal and Chemical Biomass Conversion
54. Wang S, Baxter L (2006) Fly ash and concrete. Concrete Producer 24:48–49
55. Wang S, Baxter L, Fonseca F (2008) Biomass fly ash in concrete: SEM, EDX and ESEM analysis. Fuel 87:372–379
56. Wang SZ, Baxter L (2007) Comprehensive study of biomass fly ash in concrete: Strength, microscopy, kinetics and durability. Fuel Process Technol 88:1165–1170

57. Wang SZ, Llamazos E, Baxter L, Fonseca F (2008) Durability of biomass fly ash concrete: freezing and thawing and rapid chloride permeability tests. Fuel 87:359–364
58. Wang SZ, Miller A, Llamazos E, Fonseca F, Baxter L (2008) Biomass fly ash in concrete: mixture proportioning and mechanical properties. Fuel 87:365–371
59. Kaer SK, Rosendahl LA, Baxter LL (2006) Towards a CFD-based mechanistic deposit formation model for straw-fired boilers. Fuel 85:833–848
60. Hustad JE, Sønju OK (1992) Biomass combustion in IEA countries. Biomass Bioenerg 2:239–261
61. Guo X, Bartholomew CH, Hecker WC, Baxter LL (2009) Effects of sulfate species on V2O5/TiO2 SCR catalysts in coal and biomass-fired systems. Appl Catal B Environ 92: 30–40
62. Guo X, Nackos A, Ashton J, Baxter LL, Bartholomew CH, Hecker WC (2005) Poisoning/deactivation of vanadia/titania dioxide SCR catalyst in coal and biomass fired systems. Presented at the 30th International Technical Conference on Coal Utilization and Fuel Systems. April 17–21 2005, Clearwater Coal Conference
63. Guo X, Ashton J, Butler J, Nackos A, Hecker WC, Bartholomew CH, Baxter LL (2007) Experimental Deactivation Analysis of Commercial Selective Catalytic Reduction (SCR) Catalysts. 20th North American Meeting, North American Catalysis Society
64. Guo X, Baxter LL, Bartholomew CH, Hecker WC (2005) Investigation of sulfation on vanaida/titanium dioxide catalyst. 19th North American Meeting, North American Catalysis Society
65. Guo X, Baxter LL (2006) Issues and Utilization on Biomass for Energy Conversion. 2006 American Flame Research Committee Conference

Chapter 4
Experiences on Co-firing Solid Recovered Fuels in the Coal Power Sector

Jörg Maier, Alexander Gerhardt and Gregory Dunnu

Abstract Solid Recovered Fuels (SRF) are solid fuels prepared from high calorific fractions of non-hazardous waste materials intended to be fired in existing coal power plants and industrial furnaces (CEN/TC 343, Solid Recovered Fuels, 2003). In other frameworks, these types of fuels are referred to as refuse or waste derived fuels. They are composed of variety of materials of which some, although recyclable in theory, may be in forms that made their recycling an unsound option. The use of waste as a source of energy is an integral part of waste management. As such, within the framework of the European Community's policy-objectives related to renewable energy, an approach to the effective use of wastes as sources of energy is outlined in documents like the European Waste Strategy. Within the scope of the European Demonstration Project, RECOFUEL, SRF co-combustion was demonstrated in two large-scale lignite-fired coal boilers at RWE in Germany. As a consequence of the high biogenic share of the co-combusted material, this approach can be considered beneficial following European Directive 2001/77/EC on electricity from renewable energy sources (directive). During the experimental campaigns, the share of SRF in the overall thermal input was adjusted up to 15%. The measurement campaign included boiler measurements in different locations, fuel and ash sampling and its characterization. The corrosion mechanisms and rates were analysed and monitored by dedicated corrosion probes. The scope of this chapter covers the characterisation of SRF using the pre-nominative technical specifications of CEN and the status of the standardization activities. Additionally this chapter summarizes some of the experiences gained from co-firing of SRF and biomass in large scale demonstration plants. These include handling and pretreatment of the SRF, milling corrosion, emissions behaviour, and the quality of solid residues.

J. Maier (✉)
Institute of Combustion and Power Plant Technology – IFK,
University of Stuttgart, Pfaffenwaldring 23, 70569 Stuttgart, Germany
Tel: +49 711 685 63396, Fax: +49 711 685 63491
e-mail: Joerg.Maier@ifk.uni-stuttgart.de

4.1 Background for Co-combustion

Co-firing in power plants is basically the addition of supplementary fuel(s) to the main fuel and firing both fuels simultaneous in the furnace. In most cases both fuels are solid and the main fuel is either hard coal or lignite. In the following sections the emphasis will be on co-firing of waste derived fuels.

The production and thermal utilisation of Solid Recovered Fuels (SRF) from non-hazardous bio-residues, mixed- and mono-waste streams can be a key element in an integrated waste management concept (WMC). The Community waste strategy of the European Commission (COM (96) 399 final) lays down the hierarchy of waste management policy as follows:

- first priority: *prevention of waste*;
- second priority: *recovery (material over energy)*;
- last priority: *final disposal.*

This hierarchy must be applied with certain flexibility and be guided by considering the best environmental solution taking into account economic and social costs. Where environmentally sound, preference should be given to material over energy recovery, although in certain cases preference can be given to energy recovery. Therefore the co-utilisation of SRF is enforced by the implementation of the landfill directive 1999/31/EC and promoted by the directive 2001/77/EC on electricity from renewable energy sources (RES-E). Figure 4.1 shows the elaborated hierarchy within the within the WMC.

The utilisation of SRF in energy production throughout Europe offers an enormous potential as a sustainable and environmental friendly waste-to-energy technology, whereas the high biogenic share of SRF (45–65 wt%) contributes significantly to the reduction of greenhouse gas emissions (approximately 1 Mg (megagram) CO_2 per Mg SRF) and the conservation of natural resources by substitution of fossil fuels, while the electricity costs will be significantly below 0.05 €/kWh as a major target of the Community in terms of renewable energy production.

Considering the various waste input streams used in SRF production, an urgent demand for the implementation of a sustainable quality management system can be determined, ensuring efficient and environmental friendly production and utilisation.

The EC-directive on the promotion of electricity produced from renewable energy sources in the internal electricity market (2001/77/EC) includes in its scope the production of electricity from biomass. In the context of SRF co-combustion there is a biodegradable fraction of products defined which counted as biomass, *i.e.* waste and residues from agricultural, forestry and related industries, as well as the biodegradable fraction of industrial and municipal waste. In a future harmonised market the member states must comply with current Community legislation on waste management.

Figure 4.1 The targeted hierarchy within the waste management concept [1]

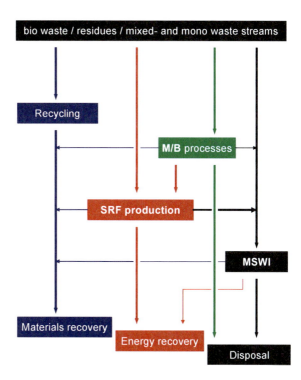

From a technical point of view, state-of-the-art, waste and residue treatment techniques of the last decades have been and still are recycling (reuse), incineration (thermal disposal) and (final) disposal on landfills. The most economic – and unfortunately in several European Countries the only treatment path – is disposal on landfills [2]. The environmental deficits (uncontrolled emissions, contamination of soils and ground water, un-recovered material or energy) are not acceptable and therefore the landfill directive was implemented.

This is one of the major driving forces to develop and implement further environmentally and economically sound alternatives in an integrated and sustainable waste management concept.

Due to liberalisation and need for cost reduction, the industry is highly interested in less expensive fuels of a specific and homogeneous quality. Recently, the main SRF users are found in the cement and lime industry, but power stations burning coal, lignite or even biomass as a primary fuel can be assessed as an emerging sector with a large potential. Further use of SRF as a carbon substitute in the steel industry is also a possibility as is mono-combustion within combined heat and power plants for district heating or process heating and electricity production.

Currently about 1.5 million tons/year of solid recovered fuel with a biogenic share of 45–65% is produced and utilised in Europe. Major countries using SRF in the European Union are Germany, Italy, the Netherlands and the Scandinavian countries.

4.2 Introduction to Solid Recovered Fuels (SRF)

SRF is produced in special waste treatment facilities operated by private and public companies. Input materials are municipal waste streams and production residues, but also packaging material (wraps), paper/cardboard and textiles. Common process technologies are:

- mechanical processing in order to separate the high calorific fraction (HCF) and to remove unwanted components (*e.g.* PVC);
- mechanical–biological treatment plants with process integrated separation and processing of HCF.

Depending on the production line, the SRF products are mainly produced as bales, fluff and soft pellets. Compared to the biomass market the production and use of hard pellets is of less interest. Waste suitable for the production of recovered solid is defined according to the waste catalogue and the Commission Decision *2000/532/EC*. According to the waste categories, the input materials can be separated in five main groups:

- Group 1: wood, paper, cardboard and cardboard boxes;
- Group 2: textiles and fibres;
- Group 3: plastics and rubber;
- Group 4: other materials (*e.g.* waste ink, used absorbers, spend activated carbon);
- Group 5: high calorific fractions from non-hazardous mixed collected wastes.

In contrast to the situation 30 years ago, the producers of SRF started the initiative for a quality monitoring system that should guarantee the properties of solid recovered fuel out of non-hazardous waste. In the past, SRF was mainly produced from process-specific wastes such as mono-batches, which were easier to handle and control. Nowadays, with increasing capacity of process technologies in terms of material identification and separation, fractions of municipal solid waste and other mixed wastes and residues play a significant role in the fuel production process and it was expected that the implementation of the landfill directive in several European Countries by 2005 (and beyond) will strengthen this development. Consequently, SRF becomes more and more a product generated out of various input streams. Although the different materials should be of non-hazardous origin, the implementation of a quality management system enabling control of the fuel input streams and the produced SRF appears to be indispensable to prevent misuse and illegal disposal, *e.g.* by dilution of critical waste streams and components in the mixed SRF.

As the need for such control mechanisms was recognised by the fuel producers, several SRF producing countries evolved quality assurance concepts like the German Institute for Quality Assurance and Certification (RAL-GZ 724), the Finnish regulation SFS 5875 and the Östereichische Gütegemeinschaft Sekundärenergieträger (ÖG SET) [3].

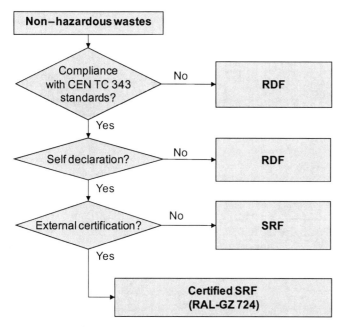

Figure 4.2   Differences regarding qualification of SRF [4]

Figure 4.2 shows the decision path when producing a certified SRF according to standardisation, declaration and certification.

4.3   European Standardisation of SRF

On the European level, which is mainly based on the final report of the CEN TASK Force 118 "Solid Recovered Fuels", the European Commission (EC) gave a mandate (M325) to CEN to develop and validate Technical Specifications (TS) for SRF, and then transform these technical specifications into European Standards (EN). The standardisation activities related to solid recovered fuels are combined and coordinated in the CEN-TC 343 and the related national mirror committees. Scope and activities of CEN/TC 343 are shown in Figure 4.3.

Interfaces are the points of waste reception with relevant specifications for the fuel producers and the point of delivery with relevant information and classification methods for the user.

With the production of classified fuels from non-hazardous waste to be used for energy recovery in waste incineration and co-incineration plants, CEN/TC 343 is dedicated to the elaboration of technical standards, specification and reports covering:

- terminology and quality management;
- fuel specification and classes;

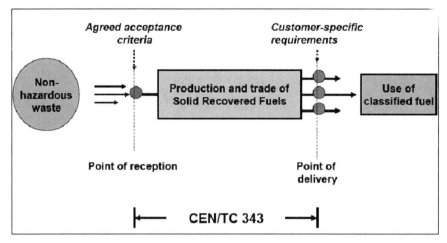

Figure 4.3 Scope and objectives of CEN/TC 343 (CEN/TS 15359 SRF-Specifications and classes)

- sampling, sample reduction and supplementary test methods;
- physical/mechanical test methods;
- chemical test methods.

4.4 Classification of SRF Within CEN TC 343

As part of the activities of CEN-TC 343 working group, a classification system for SRF has been proposed. In this instance, different SRF qualities will be categorised and a classification number assigned. It is based on three properties, the net calorific value which serves as the economic indicator, the chlorine content as the technological indicator and the mercury content as the environmental key parameter:

- net calorific value (as received) $> 3–45$ MJ/kg;
- chlorine content (as received) $< 0.1–6$ wt%;
- mercury content (as received) $< 0.02–0.5$ mg/MJ.

Table 4.1 shows the boundary conditions associated with each class number and the number of SRF associated with those classes compared to a survey made by Flamme [5] with more than 50 industrial plants in Europe. As conclusion the class limits for every regarded classification parameter were accurately chosen for the actual state and with only two exceptions an adequate class was available for each type of SRF. This means that the CEN TS 15359 (specifications and classes) is consistent with the Waste Incineration Directive.

Table 4.1 SRF classes used in European Power plants [5]

Classification property	Unit		Classes					No class
		1	2	3	4	5		
NCV	Statistical measure (Mean) [MJ/kg ar (as received)]	≥25	≥20	≥15	≥10	≥3		–
	Examination	3	10	19	25	2		1[a]
Chlorine (Cl)	Statistical measure (Mean) [% dm]	≤0.2	≤0.6	≤1.0	≤1.5	≤3.0		–
	Examination	4	18	29	10	2		1[b]
	Statistical measure (Median) [mg/MJ ar]	≤0.02	≤0.03	≤0.08	≤0.15	≤0.50		
Mercury (Hg)	Statistical measure (80th percentile)	≤0.04	≤0.06	≤0.16	≤0.30	≤1.00		
	Examination	19	11	15	8	6		

[a]SRF made from sewage sludge
[b]SRF made from municipal solid waste and sewage sludge

4.5 SRF Characterisation

Due to the recent interest in SRF, the development of European standards by CEN TC 343 for the characterisation and classifications are in the offing. This will help to distinguish it from other fuels derived from waste streams. The standards can promote the production, trade and use of solid recovered fuels and therefore support security of fuel supply in the EU. They will help permitting authorities to handle co-combustion requests from power plants and provide common procedures within the EU. With this the standards will provide methods for assessing solid recovered fuels with respect to the RES-E Directive, *i.e. "promotion of electricity produced from renewable energy sources (RES) in the internal market"*.

The clear definition of SRF in accordance with CEN TC 343 is as follow:

Solid fuel prepared from non-hazardous waste to be utilised for energy recovery in incineration and co-incineration plants and meeting the classification and specification requirements laid down in CEN/TS 13359 (Technical Specifications from the European Standardisation Committee).

Successful applications of SRF in power plants and industrial furnaces would require a thorough understanding of the fuel properties which include the combustion behaviour, emission potential, impact on facility and residues, *etc.* The determination of combustion behaviour seeks to outline possible methods and procedures that can be adopted to analyse any given SRF. An approach has therefore been outlined where the determination of combustion behaviour is categorised into groups as shown in Figure 4.4 which combine to give a holistic impression of the combustion progress of SRF in both mono and co-firing systems.

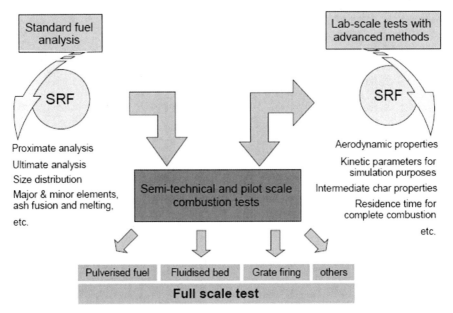

Figure 4.4   Scheme to determine combustion behaviour of SRF [6, 7]

4.5.1   *Standardised Methods*

Within the spectrum of CEN TC 343 different methods were developed and adapted to make a comparable analysis of varying SRF possible. The technical specifications cover:

- sampling and sample reduction;
- preparation of a test portion from the laboratory sample;
- calorific value;
- ash, volatile and moisture content;
- ash melting behaviour;
- bulk density, density of pellets and briquettes;
- durability of pellets and briquettes;
- bridging properties of bulk material;
- particle size and particle size distribution by screen method;
- metallic aluminium;
- digestion of material before chemical analysis;
- method for carbon (C), hydrogen (H), nitrogen (N);
- method for sulphur (S), chlorine (Cl), fluorine (F), bromine (Br);
- major elements: Al, Ca, Fe, K, Mg, Na, P, Si, Ti;
- trace elements: As, Ba, Be, Cd, Co, Cr, Cu, Hg, Mo, Mn, Ni, Pb, Sb, Se, V, Zn;
- biomass content by selective dissolution.

4.5.2 Advanced Methods

The so-called advanced methods are mostly in-house methods usually designed to provide additional information of a specific type of SRF, *e.g.* waste, or bio-residue fractions to suit a specific combustion system, *e.g.* pulverised firing system, grate firing, fluidised bed, or cement kiln. It should be mentioned that standard analyses of the SRF will determine basic parameters about the combustible and incombustible matter. The amount of energy, the contents of water, volatiles, fixed-carbon, ash, and particle size will roughly dictate the type of the combustion system that is best suited. However, vital information on the SRF with respect to the following which is needed to determine the combustion behaviour in real boilers will require additional tests:

- the burnout time of the particles;
- the aerodynamic behaviour of the particles;
- the maximum particle size required for complete combustion.

These tests are hereby grouped in the so-called advanced methods. Figures 4.5–4.7 are some dedicated laboratory methods published by several researchers in recent times.

Figure 4.5a Dedicated horizontal tube furnace for laboratory determination of SRF burnout time

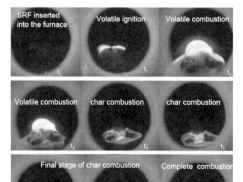

Figure 4.5b Experimental video images during burnout experiments [8]

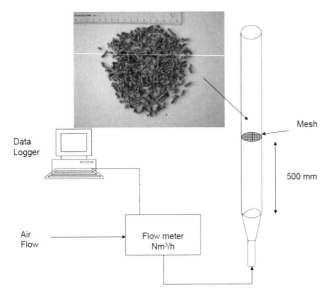

Figure 4.6 Determination of aerodynamic properties of SRF [9]

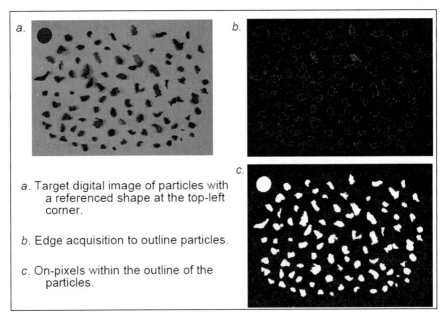

a. Target digital image of particles with a referenced shape at the top-left corner.

b. Edge acquisition to outline particles.

c. On-pixels within the outline of the particles.

Figure 4.7 Particle image analysis method (PIAM) for size analysis of SRF [9]: (a) target digital image of particles with a referenced shape at the top-left corner, (b) edge acquisition to outline particles, and (c) on-pixels within the outline of the particles

4.5.3 CFD-Simulation Tools

Beside analytical and experimental test methods the development and use of combustion simulation models to predict and improve the boiler performance is promising and should be recommended. At present there are several commercial codes based on CFD used to simulate pulverised coal boilers, *e.g.* AIOLOS, FLUENT, *etc.* The adaptation of these codes to co-firing of SRF will depend on the development and successful implementation of particle combustion models, taking into account the aerodynamics, combustion kinetics of the SRF and its char, *etc.*

Different methods and apparatus are used to determine kinetic data but the method which is applicable to pulverised fuels are somewhat not applicable to SRF, which are in the size range of 5–30 mm. What is often done is to mill the SRF to the micron range size before determination. The obvious setback in this approach is that the heat up and reactivity of pulverised SRF is totally different from the SRF fed. A new approach to deal with larger particles is therefore necessary.

The combination of standard and advanced methods in addition to bench-scale tests offers profound bases to evaluate the behaviour SRF. The acquired data are vital inputs that will improve results from sub-models which can further be implemented in CFD codes.

4.6 Industrial Scale Production and Applications of SRF

4.6.1 SRF Production

For the production of SRF some recommendations can be made and it is possible that SRF-production and MSWI can complement each other. Traditional MBT and MT are still the most widely used technology for SRF production.

Modern treatment-technology uses substantial material knowledge to produce high quality SRF with good chemical and physical fuel properties. State-of-the-art technologies, *e.g.* NIR-technology shown in Figure 4.8, make it possible to separate suitable HCFs with low Cl-values from MSW, bulky waste and mixed commercial wastes. Further developments of the NIR to improve the selectivity will assure quality management system in accordance with CEN TC 343 and RAL-GZ 724 and improve the reliability of fuel properties and the fuel-quality itself. In terms of chlorine and some alkaline earth metals content, the qualities of the SRF can be improved when implementing state-of-the-art separation technology. Data provided within the framework of RECOFUEL is presented in Figure 4.9 and shows this development for the SRF type called SBS[®]1.

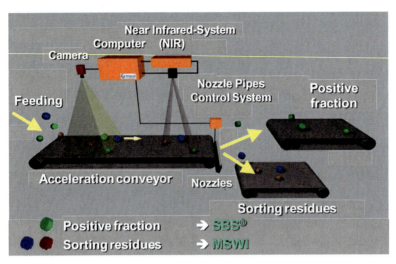

Figure 4.8   HCF-Sorting with NIR-systems positive separation modus [4]

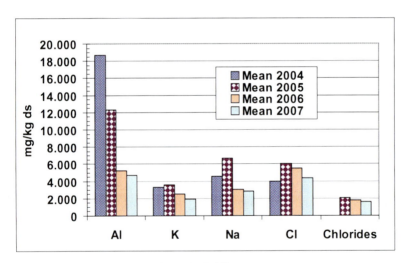

Figure 4.9   Quality assurance of SRF (SBS) [4]

4.6.2 Co-firing SRF at Coal Fired Power Plants

Due to liberalisation and the need for cost reduction, the industry is highly interested in less expensive homogeneous substitute fuels of a specific quality. Actually, the main SRF users are found in the cement and lime industry. Coal-fired
power stations can be assessed as an emerging sector with a huge potential. The

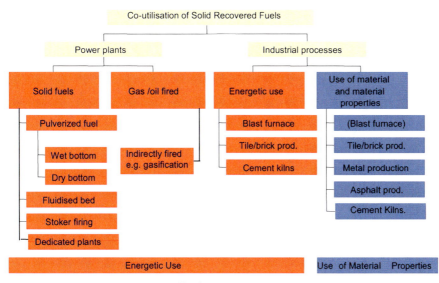

Figure 4.10 Thermal and material utilisation of SRF

steel industry uses SRF as a carbon-substitute and the Scandinavian countries burn SRF for district heating. Major countries producing SRF in the European Union are Germany, Italy, the Netherlands and Scandinavian countries. Currently about 1.5 million tons/year of SRF with a biogenic share of 45–65% is produced and utilised in Europe.

In most cases the implementation of direct co-combustion is possible without great changes and investments in existing infrastructure. It gives a cost effective option to increase the share of renewable energy. Figure 4.10 shows the different ways for using waste streams either for energetic (red) or material (blue) purposes in industrial processes.

4.6.2.1 Power Plants in Germany that Conduct Co-combustion of SRF

Currently, there are at least nine pulverised power stations in Germany – involving both lignite and hard coal – that have performed or are conducting co-firing. The following list shows the power plants and the type of supplementary fuel used for co-firing.

1. Dry bottom:

- Jaenschwalde (lignite): SRF derived from MSW, and meat-and-bone meal.
- Gersteinwerk/Werne (hard coal): SRF derived from MSW, meat-and-bone meal, and commercial waste.
- Weisweiller (lignite): SRF derived from MSW, sludges and commercial waste.

2. Wet bottom:

- Werdohl-Elverlingsen (hard coal): SRF derived from MSW, and meat-and-bone meal.
- Hamm-Westfalen (hard coal): SRF derived from MSW, and commercial waste.
- Ensdorf/Saar (hard coal): SRF derived from commercial waste, sludge, and meat-and-bone meal.

3. Circulating fluidized bed:

- Berrenrath/Ville (lignite): SRF derived from MSW, and sewage sludge.
- Flensburg (hard coal): SRF derived from MSW and commercial waste.
- Oberkirch: paper sludge and sewage sludge.

4.6.3 SRF Feeding and Pre-treatment Concept

SRF treatment for direct injection co-firing practices can basically take two forms. The first is the co-utilisation of the existing coal mills. Here the SRF and coal are milled together in the existing milling facility and the mixed fuel is fired through existing coal burners. The advantage of this configuration is its low capital investment. However, there is a limitation on how high the thermal share contribution from the SRF can go in order not to compromise the quality of the coal dust in terms of particle sizes. For higher thermal share of coal substitution, pre-treatment of the SRF, which is the second option, is preferred. Here the SRF is pre-milled in installed dedicated mills, which can be on-site or off-site. The SRF are then fired via the coal ducts or through dedicated burners. Of course this second option is capital intensive. Figure 4.11 shows a schematic representation of the options available for direct injection.

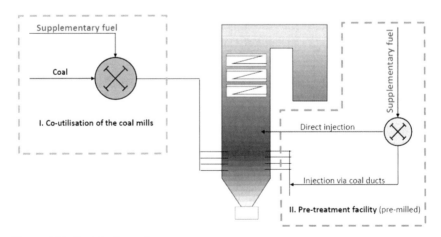

Figure 4.11 Fuel feeding and pre-treatment concepts

In pulverized fuel (PF) combustion light and fluffy fuels are normally pneumatically transported and injected into the boiler. The usual particle size distribution of coal is fine dust with 90–98% of particles smaller than 100 μm. The higher moisture content and the softness of the plastic and paper fraction result in a more problematic milling behaviour, at least by using the existing coal mills infrastructure. The maximum size for the SRF in order to reach a satisfactory burnout within the residence times typical for PF boilers is estimated to be around 10–20 mm. The average plant efficiency of the existing PF power plant is estimated to be 35–36%, although the current state of the art for the lignite fired power plants is approximately 43%.

Alternative methods for using SRF are indirect co-combustion and parallel combustion. In the indirect co-firing mode, the supplementary fuel is gasified in a separate facility and the product gas is then being fed in the coal combustion chamber. Therefore indirect co-combustion has no influence on the coal ash. For the parallel combustion concept an additional combustion facility for the SRF is needed; the separate produced steam is then fed in the coal boiler and upgraded to higher conditions. With this the energy conversion efficiency is higher and this solution may have less operational problems. An example of this concept is the UPSWING process which is explained in the following section.

For both methods, indirect co-combustion and parallel combustion, additional costs arise from the gasification or combustion facility, which make this option less favourable in most cases.

4.6.4 UPSWING Process

Another concept developed for co-utilisation of waste fuels is the UPSWING process, an acronym for "*U*nification of *P*ower *P*lant and *S*olid *W*aste *IN*cineration on the *G*rate", describing the combination of a conventional grate firing system with a power plant on both the steam and the flue gas sides. The concept was developed by the Forschungzentrum Karlsruhe, Germany, and patented 1998–2003 [10]. A schematic overview of the UPSWING process is shown in Figure 4.12,

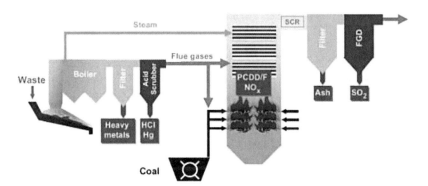

Figure 4.12 UPSWING concept (Source: Hilber, 2008 [11], originally in [12])

covering the waste-to-energy section, the partial flue gas cleaning concept as well as the integration of both flue gas and steam to the power plant.

4.6.5 Demonstration Projects

Detailed investigations and research work on the feasibility of co-firing started with solid biomass fuels in large-scale coal fired power plants in the early 1990s, for example with the Clean Coal Technology Program (APAS). In a first attempt, the activities were focussed on the preparation and feeding of the biomass fuels as well as the effects of the co-utilisation on general feasibility, environmental impact, limitations, and operational performance of the power plant.

The main focus of the first investigations has been on the co-utilisation of pure and untreated biomass types such as wood and straw. Besides untreated biomass assortments, more and more production residues, such as bark, saw dust, olive residues, *etc.*, as well as bio-waste materials, *e.g.* sewage and paper sludge, waste wood, and SRF became of interest due to their low, neutral, or even negative fuel costs.

In several R&D projects under different Frameworks (Thermie, FP5, and FP6) challenges were identified and investigated in regards to *operational and environmental problems* such as:

- boiler performance (including fouling, slagging, and corrosion);
- increased deactivation of installed SCR, DeNO$_x$ catalysts;
- reduced ESP performance;
- restriction of emissions (fine particle, HCl, Hg, *etc.*);
- quality and use of residues.

In the following sections objectives and results of some dedicated SRF co-firing demonstration projects are highlighted. All these activities are co-funded by the European Commission (DG-TREN).

4.6.5.1 Direct Co-firing Projects Co-funded by the European Commission

Within the scope of the European Demonstration Project RECOFUEL, SRF co-combustion has been successfully demonstrated in two different lignite fired boilers of RWE Power AG in Germany. A 600-MW$_{el}$ pulverised lignite-fired boiler in Weisweiler was used for a 2-week measurement campaign in 2005. The demonstration was continued at the 235-MW$_{th}$ Circulating Fluidised Bed (CFB) lignite fired boiler with combined heat power generation in Berrenrath in three phases over a 17-month period in 2007 and 2008 with two measurement campaigns in April and November 2007. During the campaigns a maximal thermal share of 4% was investigated in the pulverised fired boiler and a thermal share of 15% (approximately 6 Mg/h SRF) for the CFB. The total amount of approximately

5,000 Mg of SBS®1 (Subsitutionsbrennstoff) in the pulverised coal boiler and ca. 80,000 Mg of SBS®1 in the CFB was co-fired. The SRF production and quality control was, alongside emissions, operational behaviour, corrosion and ash disposal, one of the main topics.

Some of the observations made under this project were that [13]:

- No problems occurred concerning the combustion of SRF together with lignite.
- CFB firing systems coped with much higher thermal shares of SRF than PC firing systems.
- Design of the feeding, metering and handling systems has to account for the special demanding mechanical properties of the SRF; in other respects the installations are relative simple.
- Impurities (foreign material) cause increased wear and operating trouble, meanwhile the problem was reduced significantly during the project.
- Co-combustion of SRF with low chlorine content (SBS®1, Figure 4.8) is possible at a specific level.
- Positive effects on corrosion and ash quality could be observed by multi-fuel co-firing such as SRF and sewage sludge.
- In order to avoid high temperature chlorine corrosion, it is imperative to carry out pre-examinations and to define the required quality specifications for the SRF and the overall fuel quality regime.

The DEBCO project (started 2008) is concerned with the development and demonstration of innovative approaches to the co-utilisation of biomass and SRF with coal for large-scale electricity production and/or CHP, at more competitive costs and/or increased energy efficiency. The development, demonstration and evaluation of innovative and advanced co-firing technologies will arise. The implications of the findings from the work, for future co-firing projects, involving both the retrofit of existing pf-coal-fired power plants and the provision of advanced co-firing capabilities in new-build projects, will be assessed.

Under the DEBCO project, it is proposed to perform a number of demonstrations of relevant biomass and SRF co-firing technologies with long-term monitoring and assessment of the key technical aspects, viz:

- the fuel supply chains (local available and international treated fuels);
- the bio-fuel qualities (agriculture residues, energy crops, wood pellets, *etc.*);
- the application of advanced co-firing techniques (50% and more) to a number of pulverised fuel power plants burning both lignite and bituminous coals;
- the detailed evaluation of the role of co-firing in a sustainable energy market, including both the technical and socio-economic impacts.

Under this demonstration project, Fusina Power plant in Italy has co-fired RDF with coal up to 2.5% thermal share in a 320-MWe boiler (units 3 and 4) during the period 2004–2008, and subsequently increased the thermal share to 5% since April 2009 [14].

Some of the critical issues raised during coal/RDF co-firing are in the area of RDF milling process, the feeding and on the boiler. The RDF caused high wear on the mill blades, the grates, and also on the transporting pipes due to the presence of metals and inert materials, *e.g.* glass. On the boiler side, it was reported that a significant increase of bottom ash amounts was observed. Also, an increased boiler slagging was observed but the originally designed soot blowing system was normally effective without an additional revamping system. Nevertheless some plugging problems were observed on the boilers with some coal/RDF eutectic mixtures that led to unscheduled outages. Despite the high RDF chlorine content (0.9% upper limit), corrosion problems were not yet experienced in the boiler and in the cold regions. Moreover no significant impact of RDF co-firing on NO_x and CO concentrations in the flue gas at the boiler outlet and unburned carbon in fly ash has been noticed [14].

4.6.5.2 Indirect Co-firing: Polygeneration Through Gasification Utilising Secondary Fuels Derived from MSW

Extensive development and demonstration activities have also been performed for different biomass, SRF pre-treatment systems, *e.g.* washing, pyrolysis and gasification, such as Amer, Zeltweg, Lahti 1 projects, with varying success at the commercial level. Furthermore, results and experiences from these projects showed that the dimensioning, design and implementation of required gas cleaning devices of such pre-treatment plants must be adjusted carefully to the individual boundary conditions at the power plant site.

In the city of Osnabrück, Germany, Herhof GmbH operates, under its patented mechanical biological drying technology, a 90,000 tons/year Municipal Solid Wastes (MSW) recycling plant, where approximately 50% of incoming MSW (45,000 tons/year) is converted into a secondary fuel (marketed under the name Stabilat), currently used in cement plants. A portion of Stabilat of low quality (approximately 500 tons/h) will be converted to electricity and heat in a gasification plant in order to provide the energy needs for the plant and also generate a high quality recovered fuel. The syngas will be combusted in a gas boiler and the steam generated to run a 0.5 MWe steam turbine. The electricity produced will be fed to the grid, while waste heat will be utilised in the recycling plant [15].

4.7 Summary

Experience in the field of co-combustion of different SRF in coal fired boiler shows that this technology is feasible and that all challenges can be solved with the technical and scientific resources available.

The contribution to an integrated waste management concept and, with the biogenic content in mind, to the reduction of fossil CO_2 makes co-combustion an even more sustainable and highly economic option.

As highlighted in many articles, technical issues need to be addressed individually according to the boundary conditions at each plant. The fuel handling, storage, grinding and feeding are one of the key parameters for a successful long-term operation.

To monitor and control on a long-term basis the fuel production and supply chain standardised methods are essential and, according to the effort spent in the CEN-TC-343 ,such methods are now available as technical specifications.

Besides standard methods, good communication between user and supplier is recommended to avoid operational or environmental problems and both will substantially improve the overall availability of the production and utilisation plants.

Besides technical, environmental and economical advantages of co-firing, this technology is able to explore on a short-term basis a high market potential which forces the implementation of an efficient production and logistic system.

With respect to the reduction of greenhouse gases, co-firing even short-term in highly efficient power plants could support substantially the goals set by the governments.

According to valuable experience in Europe with this technology, supported by many European co-funded research and demonstration projects, the European manufacturers, suppliers and utilities are leading competitors on the international market.

References

1. Gawlik BM, Ciceri G (2005) Quovadis Waste-to-Fuel Conversion ISBN92-8949861
2. European Commission Directorate General for Environment (2003) Refuse derived fuels-Current situation and perspectives. (B4-3040/2000/306517/MAR/E3) Final report
3. Maier J, Hilber T, Scheffknecht G (2005) Current activities in terms of Solid Recovered Fuel (SRF) standardisation. Wastes and Biomass Thermal Treatment Combustion and Co-Combustion, 16–17 November, Wroclaw, Poland ISBN: 83-7085-908-9
4. Glorius T (2008) RECOFUEL 8th Technical meeting. Hürth
5. Flamme S (2008) Main conclusions of the EU-project Quovadis. RECOFUEL Workshop Hürth, Germany
6. CEN/TR 15716: (2008) Solid recovered fuels – determination of combustion behaviour
7. Maier J, Dunnu G, Gerhardt A, Scheffknecht G (2008) Role and development of coincineration in a sustainable waste management strategy. 33rd International Technical Conference on Coal Utilization and Fuel Systems, Clearwater USA
8. Dunnu G, Maier J, Hilber T, Scheffknecht G (2009) Characterisation of large solid recovered fuel particles for direct co-firing in large PF power plants. J Fuel, doi:10.1016/j.fuel.2009.03.004
9. Dunnu G, Hilber T, Schnell U (2006) Advanced size measurements and aerodynamic classification of solid recovered fuel particles. Energ Fuel 20(4):1685–1690
10. Hunsinger H, Kreisz S, Seifert H, Vehlow J (1998) Verfahren zur Beschickung der Verbrennungseinheit eines Kohlekraftwerks
11. Hilber T (2008) UPSWING: an advanced waste treatment concept compared to state-of-the-art. Cuvillier Verlag Göttingen ISBN 978-3-86727-563-7
12. Vehlow J, Hunsinger H, Kreisz S et al. (2003) UPSWING – a novel concept to reduce costs without changing the environmental standards of waste combustion. IEA Bioenergy Joint Task Seminar, Tokyo

13. Maier J, Gerhardt A, Röper B *et al.* (2009) Mitverbrennung von Sekundärbrennstoffen in Feuerungen mit rheinischer Braunkohle. VGB, Kassel
14. Rossi N, La Marca C, Lattanzi S (2009) RDF co-firing at Enel Fusina power plant. 17th European Biomass Conference and Exhibition, Hamburg 29th Jun–2nd Jul
15. POLY-STABILAT, EU project under the 7th Framework, No TREN/FP7EN/219062

Chapter 5
Biomass Combustion Characteristics and Implications for Renewable Energy

Hong Lu and Larry L. Baxter

Abstract Unlike pulverized coal, biomass particles are neither small enough to neglect internal temperature gradients nor equant enough to model as spheres. Experimental and theoretical investigations indicate particle shape and size influence biomass particle dynamics, including essentially all aspects of combustion such as drying, heating, and reaction. This chapter theoretically and experimentally illustrates how these effects impact particle conversion. Experimental samples include disc/flake-like, cylindrical/cylinder-like, and equant (nearly spherical) shapes of wood particles with similar particle masses and volumes but different surface areas. Small samples (320 μm) passed through a laboratory entrained-flow reactor in a nitrogen atmosphere and a maximum reactor wall temperature of 1,600 K. Large samples were suspended in the center of a single-particle reactor. Experimental data indicate that equant particles react more slowly than other shapes, with the difference becoming more significant as particle mass or aspect ratio increases and reaching a factor of two or more for particles with sizes over 10 mm. A one-dimensional, time-dependent particle model simulates the rapid pyrolysis process of particles with different shapes. The model characterizes particles in three basic shapes (sphere, cylinder, and flat plate). With the particle geometric information (particle aspect ratio, volume, and surface area) included, this model simulates the devolatilization process of biomass particles of any shape. Model simulations of the three shapes show satisfactory agreement with the experimental data. Model predictions show that both particle shape and size affect the product yield distribution. Near-spherical particles exhibit lower volatile and higher tar yields relative to aspherical particles with the same mass under similar conditions. Volatile yields decrease with increasing particle size for particles of all shapes. Assuming spherical or isothermal conditions for biomass particles leads to large errors at most biomass particle sizes of practical interest.

L. Baxter (✉)
Brigham Young University,
Provo, UT 84601, USA
e-mail: larry_baxter@byu.edu

5.1 Introduction

Biomass is a CO_2-neutral energy source. An attractive near-term option for utilizing biomass is cofiring biomass with coal in existing coal-fired utility boilers [1–5]. However, there are fundamental physical and chemical differences between coal and biomass that lead to important differences in combustion behavior. Biomass particles are typically much larger than pulverized coal particles because biomass preprocessing to the same sizes as typical pulverized coal particles is neither necessary nor economical. The average pulverized coal particle size is ~50 μm with top sizes of 100–120 μm, whereas a biomass particle can be up to 200 times as large. Biomass also has a much greater volatile content and often much higher moisture level than coal. However, biomass particles have considerably lower densities than coal particles, commonly differing by a factor of 4–7. Considering all of these factors, the drying, devolatilization, and oxidation time scales of millimeter-sized biomass particles typically exceeds that of pulverized coal particles under similar conditions. In particular, the resistance to intra-particle heat and mass transfer is likely to be important for biomass particles. Particle size is a major factor in determining the role of transport limitations during reaction [6]. Theoretical frameworks for incorporating intra-particle heat and/or intra-particle mass transport for large coal particles in fluidized bed combustors and gasifiers appear in the literature [6–8]. It is generally agreed that devolatilization of small coal particles (< 100 μm) occurs with negligible internal temperature gradients, mainly because the particles are so small that the gradients lead to small temperature differences. The time scale of diffusion and bulk flow of gaseous volatile species in 100-μm coal particles is ~ 10–3 s [9]. For large coal particles (> 1 mm), thermal and transport limitations dominate compared to reaction time [6]. Based on this discussion, intra-particle heat and mass transfer effects are likely to be important in devolatilization of millimeter-sized biomass particles. The large particle size, high volatile and moisture content, and possible transport-limited devolatilization characteristics of biomass particles may influence heat release, pollutant generation, carbon conversion, boiler efficiency, ash deposition [10], and fouling in the reactor [2, 11]. Several papers have examined the effects of intra-particle heat and mass transfer on biomass heating and devolatilization. Kanury [12] presents the transport equations for heating, devolatilization, and combustion of biomass considering temperature gradients within the particle. Saastamoinen examined intra-particle effects on heating, drying, and devolatilization of large wood logs. Several researchers [13–15] have theoretically examined the effects of intra-particle heat and mass transfer for large biomass particles (centimeter-sized) when subject to convection and radiation heat transfer at the surface. However, it is difficult to draw conclusions from this previous research to the conditions found in utility power generation systems for millimeter-sized particles with rapid heating rates and high temperatures. CFD-based simulations of pc boilers have also attempted to model devolatilization and combustion of coal and biomass particles [16–18]. These CFD models typically employ a relatively simple framework for predicting coal/biomass particle devolatilization and combustion that treats the particle with a lumped-sum heat capacity, similar to the treatment of coal particles. Heating, drying, devolatilization, and char combustion can all occur in parallel in most codes, but the single particle temperature associated with the typical model treatment generally leads to the processes occurring sequentially as the particle flows through the reactor. This simple single-temperature approach appears to be adequate to describe pulverized coal in most cases, but may not be a realistic model for biomass particles because of the large particle size and high volatile/moisture contents. This discussion experimentally illustrates the effects of both size and shape on biomass particle conversion mechanisms, rates, and yields, and mathematical models that capture these effects.

5.2 Distinguishing Biomass Combustion Characteristics

Coal represents the dominant low-grade, ash-forming fuel by any measure – energy generation, mass consumption, CO_2 production, and economic influence. Consequently, a great many of the experimental techniques and theoretical approaches used for many low-grade fuels, including biomass, are usefully considered in comparison to coal. This section discusses biomass broadly, with most of the discussion pertaining to traditional biomass (forest products, agricultural products, and residues). Black liquor represents another important class of biomass but, while black liquor is a form of biomass and shares many of the characteristics discussed here with other biomass fuels, it is unique among even biomass fuels in important ways. A more complete discussion involving similar experimental and modeling approaches as are considered here applied to black liquor combustion appears elsewhere [19]. Similarly, this discussion focuses on biomass particle combustion. Biomass influence on corrosion, flame structure, pollutant emissions, fly ash properties and use in concrete, catalytic systems, CO_2 emissions, economics, *etc.* all represent important additional issues that are discussed elsewhere [4, 20–37].

Table 5.1 summarizes the chemical compositions typical of several important classes of fuels. A plot of atomic ratios of hydrogen and oxygen to carbon (Fig-

Table 5.1 Average ultimate analysis results of many thousand coal and biomass samples

	Anthracite	Bituminous	Sub-bituminous	Lignite	Grass	Straw	Wood chips	Waste wood
C	90.22	78.35	56.11	42.59	45.34	48.31	51.59	49.62
H	2.85	5.75	6.62	7.40	5.82	5.85	6.14	6.34
N	0.93	1.56	1.10	0.73	2.04	0.78	0.61	1.01
O	5.03	11.89	35.31	48.02	45.95	44.18	41.57	42.89
S	0.96	2.43	0.84	1.15	0.24	0.18	0.07	0.07
Cl	0.03	0.08	0.01	0.01	0.62	0.70	0.02	0.06

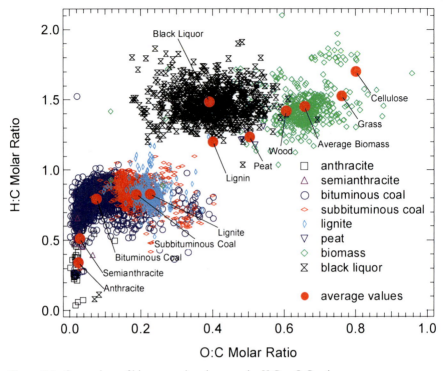

Figure 5.1 Comparison of biomass and coal on a molar H:C *vs* O:C ratio

ure 5.1) reveals a useful relationship among these fuels. All individual data points in this figure represent sample results whereas the round, red, labeled data represent averages for the samples in each classification based on averages of many hundreds to several thousands of individual measured results.

Results of classical analyses of biomass and coal fuels reflect many, but not all, of the important differences in these fuels. Table 5.1 summarizes data from coal and biomass databases including several thousands of samples of fuels in various categories. As is seen, biomass differs from coal primarily in its higher oxygen, lower carbon, and lower sulfur content. Herbaceous biomass (grasses and straws) also contain much more chlorine as a class than do either coal or wood. Some of these differences appear in Figure 5.1, where coal, biomass, black liquor, and a few other fuel compositions appear on a molar H:C to O:C ratio. Other important characteristics of biomass relative to coal include more widely ranging (0.1–25%) ash contents, lower heating values (by about half on an as-delivered basis), much lower bulk energy densities (by more than an order of magnitude), generally higher alkali content for herbaceous biomass dominated by potassium, and substantially more fibrous and less friable character. Each of these characteristics lends specific characteristics to the fuels, only a few of which appear in the following discussion.

5.3 Particle Size

Traditional biomass particle sizes of commercial interest typically range from 2 to 10 mm. This exceeds, for example, pulverized coal particle sizes and volumes by orders of magnitude. Biomass particles also exhibit highly aspherical characteristics and in this section size is characterized as the sphere-equivalent diameter, that is, the diameter of a sphere that has the same volume/mass of fuel as the biomass particle. There are several implications of this large size on biomass combustion, especially when compared to traditional coal combustion concepts. The focus here is on issues that impact practical system operation.

Figure 5.2 illustrates visible-light, temperature-map, and emissivity-map images of an approximately 2 mm diameter burning biomass (black liquor) particle/droplet [38, 39] during both devolatilization (top) and char burning (bottom). Similar images for many biomass particles appear in this chapter, but this black liquor image is useful to highlight some of the results. The devolatilization image illustrates both the particle (roughly spherical shape at center and abnormal shape in lower left) and soot (bright glow above and below the spherical shape) combustion. The temperature map indicates the large range of temperatures existing on the particle surface, with surface temperatures varying by several hundred degrees, highest in the most exposed regions of the particle surface (lower left) and lowest in the regions most sheltered from the upward moving convective flow in this

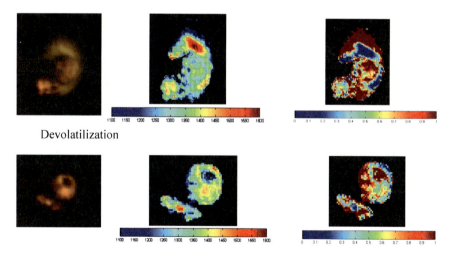

Devolatilization

Char burning (0.26 s later)

Figure 5.2 Visible-light images (*left*), temperature maps (*center*) and emissivity/emittance maps (*right*) of approximately 2 mm diameter burning biomass (black liquor) particle/droplet during devolatilization (*top*) and oxidation (*bottom*)

experiment (top and left of particle). The soot, on the other hand, exhibits much higher temperatures than those on any portion of the particle surface, in part because the overall particle temperature is still rapidly rising in these conditions and in part because the soot particles are burning under conditions near the peak adiabatic flame temperature for this system. The soot clouds, however, are relatively thin and hence have very low emissivity or emittance (right diagram).

During the early stages of oxidation illustrated in the second series of photographs, the soot cloud is gone and the particle temperatures are on average higher than during devolatilization, although there remain significant temperature differences on the particle surface. This image of the same particle is particularly instructive, however, because the particle has developed a blow hole in the surface into which oxygen cannot effectively penetrate. The particle surface temperature at this early stage of oxidation exceeds the interior temperature by at least 300 K, illustrating the very large internal temperature gradients in the particle. The emissivity map illustrates how the particle emissivity varies from near unity in this blow hole (which acts somewhat like a cavity radiator) and in regions of high char concentration to as low as 0.2 in regions where the very high sodium-salt concentrations in the char particle dominate the emittance.

These large radial and surface variations in particle properties do not exist during pulverized coal combustion. They lead to significant overlaps in particle drying, devolatilization, and oxidation processes and considerably complicate computer modeling of conversion processes. Additionally, larger particles have large boundary layers in which particle off gases, such as CO, react with bulk gas components, such as O_2, providing significant thermal feedback to the particles. These near-particle flames, which do not occur in pulverized coal combustion, strongly influence the rates of particle heating and reaction.

5.4 Particle Shape

Independently of particle size, the unusual shapes of biomass particles also lead to more complex particle behaviors than commonly occur during coal or other equant particle combustion. Pulverized coal particles approximate spheres, with aspect ratios rarely exceeding 2. Biomass particle aspect ratios commonly exceed 6 and shapes more commonly resemble cylinders or plates/flakes than spheres. A sphere exhibits the smallest surface-area-to-volume ratio of any shape and therefore represents an extreme case in processes where such a ratio is important. Large particles commonly burn at or near diffusion-limited rates, which makes this ratio of paramount importance in predicting overall conversion times. Spheres, therefore, generally represent poor approximations for many biomass particles in combustion models.

Figure 5.3 illustrates the impact of particle shape on biomass combustion characteristics generally. This figure compares the measured and predicted mass loss

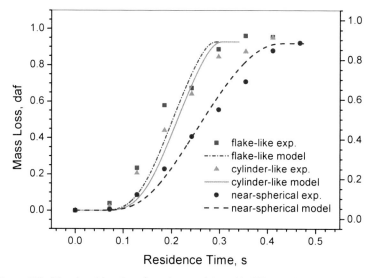

Figure 5.3   Mass loss histories of sawdust particles with different shapes

profiles of relatively small (300 µm wood) biomass particles hand-sorted into different shapes. As indicated, the near-spherical biomass particles combust more slowly than the higher aspect ratio particles. Since both heat and mass transfer rates mainly depend on surface area for these large particles, the lower surface-area-to-mass ratios of spheres significantly decrease the overall combustion rate of these particles relative to less symmetric particles with the same total mass. Detailed discussions of the model and data on which these figures are based appear elsewhere [40].

Experimental constraints limited these measurements illustrated in Figure 5.3 to relatively small particles. At more realistic biomass particle sizes, the impact on particle conversion time becomes increasingly dramatic. Figure 5.4 illustrates the predicted ratio in overall conversion time for particles of various shapes as a function of sphere-equivalent diameter, with all times normalized to those of flakes. In the size range of greatest interest to most commercial biomass conversion facilities, the conversion times of the aspherical biomass particles change by factors of 2–3 relative to that of the spheres with the same mass. As indicated in the figure, this impact becomes increasingly less significant as particle size decreases, becoming relatively unimportant for pulverized coal particle sizes (100 µm and less). The data included in this figure summarize measurements such as those illustrated in Figure 5.3 but at many different particle sizes. Many additional data confirm the predictions of Figure 5.4 over a broader size range. These differences become even more pronounced as heating rate increases from the relatively modest rates used here.

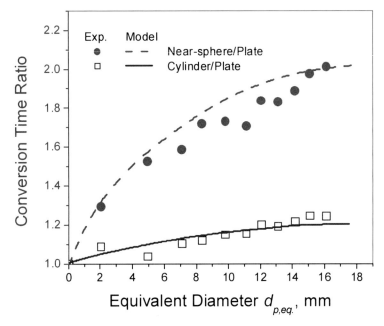

Figure 5.4 Conversion time *vs* particle sphere-equivalent diameter for particles with aspect ratios of 5 (particles in inert – N_2 – environments with 6% initial moisture with gas and radiative temperatures of 1050 K and 1273 K, respectively). At higher aspect ratios, the ratio of conversion time increases (rises to approximately 3 for aspect ratios of 8 – which is common in biomass fuels)

5.5 Moisture and Volatiles Contents

The shape, size, and relatively high moisture and volatiles contents of biomass particles influence conversion histories in practical ways. Figure 5.5 illustrates predicted centerline and surface particle temperature histories compared with measured particle temperatures. The interior data come from a thermocouple imbedded in an approximately 9.5 mm initial diameter biomass particle. Model comparisons for similar data from biomass particles/droplets ranging from < 1 to > 10 mm and a detailed description of the model appear in the literature [37, 40, 41].

The figure indicates that the center of the droplet remains wet and vaporizing well after the surface reaches devolatilization temperatures, as indicated by the low and sub-boiling predicted center temperature coinciding in time with surface temperatures over 700 K. Biomass particles begin devolatilization and swelling at lower temperatures than do coal particles, mostly due to the overall lower molecular weight of their condensed-phased organic constituents. Heat conduction along the thermocouple leads will bias the early temperature data in Experiments 1 and 2 to high values – a problem largely avoided in Experiments 3 and 4 by running the leads along the axial rather than the radial direction of these cylindrical particles.

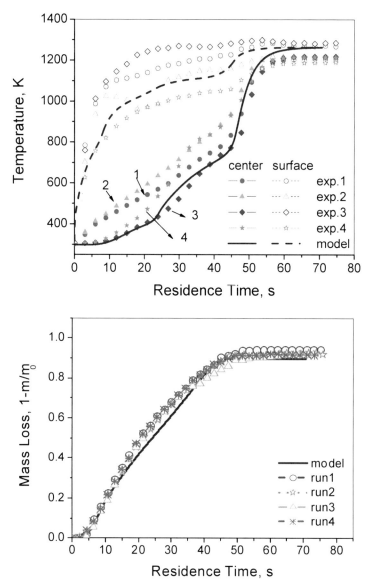

Figure 5.5 Comparison of data and model predictions of an initial 9.5-mm droplet in a 1050 K inert gas (N_2)/1273 K radiative environment furnace in nitrogen (aspect ratio = 5, initial moisture 6%, wood particle). The early temperature data in experiments 1 and 2 are biased because of conduction along thermocouple leads

Larger droplets/particles and higher heating rates (these data come from a 1050 K gas/1273 radiative environments in an inert N_2 environment) generate larger temperature gradients still. Predicted composition plots confirm this inference from the temperature, but we have yet to devise means of generating anything other than overall mass loss with which to confirm such predictions. The interior temperature measured by the thermocouple generally lies between the predicted surface and center temperatures as would be expected. The measured temperature overshoot centered around 5 s residence time may arise from some residual oxygen in the nominally nitrogen-purged furnace in which these experiments occurred, or it may relate to exothermic devolatilization reactions that occur during the latter stages of devolatilization.

In addition to the water-vaporization-related temperature plateau seen at the particle center, the rapid volatile loss also impacts temperature rise. Under the relatively mild conditions of these experiments, this impact is subtle and is largely responsible for the departure of the observed temperature rises from smooth exponential-like profiles that would exist in the absence of such blowing effects. However, under more severe conditions, the volatile yields and especially the combustion of such volatiles in the particle boundary layers have pronounced impacts on particle temperature histories.

These processes combine to produce particle off gas compositions and particle reaction timelines that differ substantially from those that would be predicted if biomass particles were treated as large, low-density, coal particles using traditional modeling techniques.

There are additional considerations associated with transformations of particle inorganic material into fly ash and its subsequent impact on deposition, corrosion, SCR performance, and fly ash utilization. These considerations will appear in other documents.

5.6 Flame Proximity to Particle

Biomass particles, by virtue of their size and, more specifically, the size of their boundary layers, commonly include part or all of the flame within the thermal and mass-transfer boundary layer. This differs from coal particles, which can accurately be described with single-film models. The effects of this flame in the boundary layer generally include: (1) an increase in particle surface temperature during most of the oxidation stage of combustion of about 100°C, (2) more uniform particle temperatures, (3) changes in boundary layer thickness. However, most of these influences occur only for small fractions of the total reaction time, occur mostly during oxidation, which for biomass represents a small fraction of the total mass, and occur near the end of the particle lifetime where they have no cascading effects.

5.7 Single Particle Combustion Model

The details of the single biomass particle combustion model used in these analyses appear elsewhere and are not repeated here [37, 42]. In short, the model describes a

transient, multi-dimensional, aspherical particle during drying, pyrolysis, oxidation, inorganic reactions, and fly ash formation for arbitrary size, shape, and composition. The following section compares these model predictions with measured
data. As discussed, there are regions of unreliable data, especially where intrusive
probes track internal temperatures. The model predictions are more reliable than
the data in these regions. Otherwise, the model and data present a consistent picture
of the effects of particle shape, size, and composition on reaction, as is discussed in
the next section in detail and as has already been summarized in the introduction.

5.8 Particle Combustion Experiments

5.8.1 Particle Devolatilization

Data for a near-spherical particle ($d_p = 11$ mm) with aspect ratio of 1.0 and a moisture content of 6.0 wt%, including mass loss, center and surface temperature during pyrolysis, appear with model predictions in Figures 5.6 and 5.7. The nominal
conditions of this experiment include a reactor wall temperature of 1273 K and gas
temperature of 1050 K. All the following validation experiments involved the
same conditions.

The particle mass loss and particle surface temperature predictions generally
agree with experimental data except that the measured particle center temperature

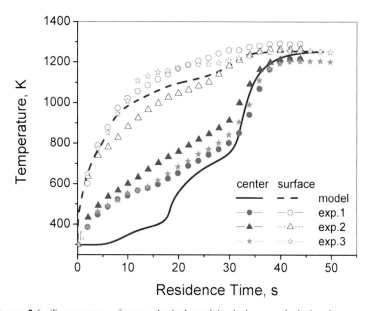

Figure 5.6 Temperature of near-spherical particle during pyrolysis in nitrogen: $d_p = 11$ mm,
$AR = 1.0$, $MC = 6$ wt%, $T_w = 1276$ K, $T_g = 1050$ K

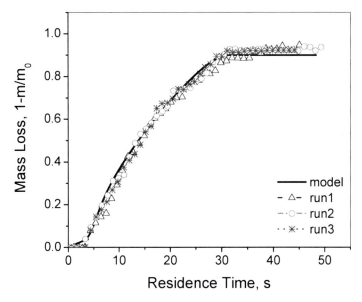

Figure 5.7 Mass loss of near-spherical particles during pyrolysis in nitrogen: $d_p = 11$ mm, $AR = 1.0$, $MC = 6$ wt%, $T_w = 1276$ K, $T_g = 1050$ K

increases faster than the model predictions at the beginning. This might be caused by the thermal conduction effects through the wire while measuring the particle center temperature. In principle, the measured particle surface temperature and center temperature should reach the same value at the end of pyrolysis, but a small discrepancy exists due to reactor temperature non-uniformity and differences in thermocouple bead size and shape. A more detailed discussion of the features of these data appears after discussing the potential cause of the discrepancy in the center temperature data.

To determine the thermocouple lead wire impact on the measured center temperature, a second experiment at the same conditions used a cylindrical particle of the same diameter but with aspect ratio of 4.0. Two thermocouples monitored the center temperature, one passing axially and a second passing radially through the particle. The axial thermocouple should be less impacted by heat conduction through the leads since the particle provides some insulation from the radiation and bulk-flow convection. In Figure 5.8, lines 1 and 2 are particle center temperatures measured in the radial direction, and lines 3 and 4 are results measured in axial direction. As indicated, the center temperature measured in the radial direction increases much faster than that measured in the axial direction at the beginning, indicating that the thermocouple wire conduction influences initial center temperature measurements. The model prediction for the center temperature generally agrees with the average of the axial direction.

Mass loss data collected in several runs for the cylindrical particle are compared with model predictions in Figure 5.9.

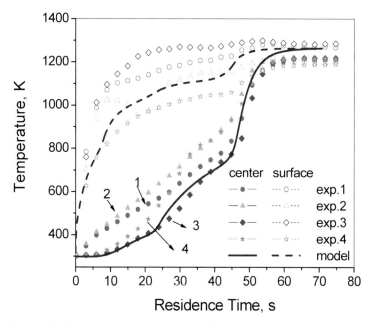

Figure 5.8 Temperature comparison of a cylindrical particle during pyrolysis in nitrogen: lines 1 and 2 indicate radial thermocouple results and lines 3 and 4 represent axial thermocouple results, $d_p = 11$ mm, $AR = 4.0$, $MC = 6$ wt%, $T_w = 1276$ K, $T_g = 1050$ K

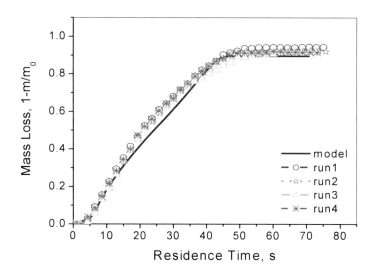

Figure 5.9 Mass loss comparison of a cylindrical particle during pyrolysis in nitrogen: $d_p = 11$ mm, $AR = 4.0$, $MC = 6$ wt%, $T_w = 1276$ K, $T_g = 1050$ K

The shapes of the temperature histories illustrate the complexity of this large-particle pyrolysis process even in the absence of complications arising from surface oxidation and surrounding flames. The initial low center temperature is associated with vaporization, which occurs at sub-boiling temperatures under nearly all conditions. Experiments with moister particles reported later illustrate more clearly the impacts of vaporization. After vaporization, particles heat up relatively slowly, mainly because of devolatilization reactions in outer layers of the particle generate significant gas velocities in the pores (commonly reaching 0.2 m/s), thereby impeding internal heat transfer. After devolatilization, the rate of particle heating increases rapidly, mainly because the particle mass is greatly reduced relative to the early data by virtue of volatile losses but significantly because the internal heat transfer impediment from rapid outgassing also subsides. By contrast, the surface particle temperature increases rapidly and is less susceptible to slow heat transfer rates or even significant impacts from the blowing factor, in this case because radiation is the dominant heating mechanism. If convection were the primary heating mechanism, surface temperature heating rates would decrease by factors of up to 10 during rapid mass loss. These processes result in temperature differences between the surface and the center of many hundreds of degrees Kelvin during particle heat up.

5.8.2 Particle Drying

The drying model was further tested using wet particles with higher moisture content. Particle surface temperature and center temperature were measured with type K thermocouples in a cylinder particle with 40 wt% moisture (based on total wet particle mass) during drying and devolatilization. Similar to the previous experiments, particle center temperature was measured in both axial and radial directions. Results appear in Figure 5.10, which illustrates model predictions compared to data. Lines 1 and 2 indicate the center temperature measured in the radial direction and lines 3 and 4 indicate the axial measurement. Both the model prediction and experimental data showed that the particle temperature first rises to a constant value near but below the boiling point, with evaporation mainly occurring in this stage. Following drying, the particle temperature again increases until biomass devolatilization slows the particle heating rate due to endothermic decomposition of biomass materials (minor impact) and the effect of rapid mass loss on the heat transfer coefficient – often called the blowing parameter (major impact). Once all biomass material converts to char, light gas, and tar, the residual char undergoes a rapid temperature rise due to its lower mass (major impact), lower heat capacity (minor effect), and return of the blowing factor to near 1 (major effect). During most of the particle history, the predicted surface temperature is approximately 200 K below the average measured surface temperature. The predicted surface temperature depends primarily on radiative heating, convective heating, the impact of the blowing factor on heat transfer, and the rate and thermodynamics of water vaporization. As discussed later, the blowing factor in this radiation-dominated environment has little impact on the predictions. The thermodynamics of water vaporization are in little doubt, although the thermodynamics of the chemically adsorbed water losses are relatively uncertain. It is also possible that the reactions of the particle with its attendant changes in size and composition compromise the thermal contact between the surface thermocouple and the particle. There is no clear indication of whether the discrepancy arises from experimental artifacts or from uncertainties in emissivity and transport coefficients or other factors.

Figure 5.11 compares the predicted and measured mass loss data. The model does not predict the measured trend within its uncertainty though the predictions and measurements are in qualitative agreement. The disagreement is likely related to the temperature issues discussed above, including the non-uniformity of reactor temperature distribution. For a cylindrical particle horizontally oriented in the center of the reactor, its ends were exposed to a higher temperature environment but the model applied at an average bulk gas center temperature.

The model was also evaluated with wet near-spherical particle drying and devolatilization data, as illustrated in Figures 5.12 and 5.13. Results show that the predicted mass loss curve agrees well with experimental data. Both the surface temperature and center temperature profiles are similar to those for the wet cylinder particle illustrated above. The surface temperature data show that the particle surface temperature rises to the water sub-boiling point and, presumably after the surface dries, rises rapidly. The center temperature data show qualitative behavior similar to that of the cylinder except that the impacts of heat conduction in the thermocouple leads remain in all of the data.

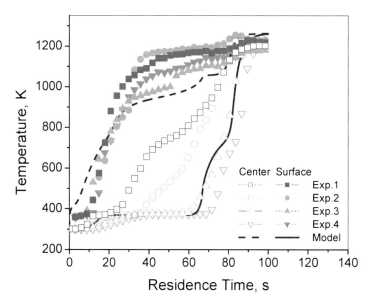

Figure 5.10 Temperature comparisons of a cylindrical particle during drying and pyrolysis in nitrogen: $d_p = 11$ mm, $AR = 4.0$, $MC = 40$ wt%, $T_w = 1276$ K, $T_g = 1050$ K

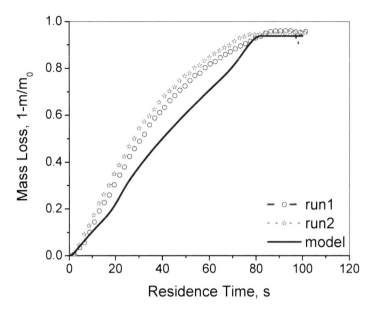

Figure 5.11 Mass loss of a cylindrical particle during drying and pyrolysis in nitrogen: $d_p = 11$ mm, $AR = 4.0$, $MC = 40$ wt%, $T_w = 1276$ K, $T_g = 1050$ K

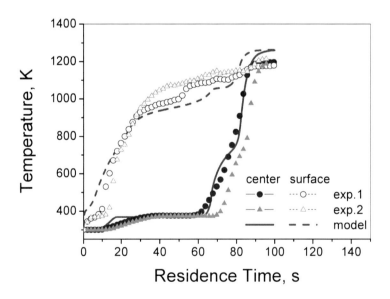

Figure 5.12 Temperature data of a wet near-spherical poplar particle during drying and pyrolysis in nitrogen: $d_p = 11$ mm, $AR = 1.0$, $MC = 40$ wt%, $T_w = 1276$ K, $T_g = 1050$ K

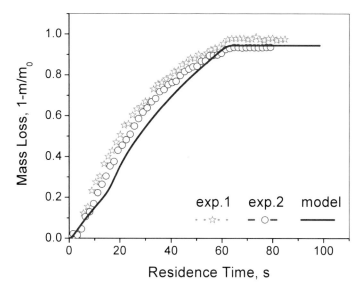

Figure 5.13 Mass loss of a wet near-spherical poplar particle during drying and pyrolysis in nitrogen: $d_p = 11$ mm, $AR = 1.0$, $MC = 40$ wt%, $T_w = 1276$ K, $T_g = 1050$ K

Figure 5.12 indicates measured surface and center temperatures increase more rapidly than the predicted values. The surface temperature discrepancy among the experimental data and model result can be explained by the experimental setup. When the wet particle was inserted into the reactor, the particle surface started to dry up. The thermocouple could not measure exactly the surface temperature since it was buried right next to the surface, which stayed near the boiling point as shown in the data. After the particle dried to some extent, the particle started to shrink and/or crack, compromising the contact efficiency of the thermocouple. On the other hand, for the center temperature measurement discrepancy, the heat conducted from the hot environment to the thermocouple bead through the thermocouple wire may still be the major influence in the temperature measurements when the particle is small as illustrated above for the dry cylinder particle.

5.8.3 Particle Combustion

Figure 5.14 illustrates the temperature profiles of a wet, near-spherical particle with 40 wt% moisture content (based on the total wet particle mass) and aspect ratio of 1.0 during the combustion process.

A type B thermocouple provides temperature data for combustion experiments since the peak temperatures exceed the reliable range of type K thermocouples. The measured particle surface temperatures are not consistent with model prediction due to experimental artifacts associated with a shrinking particle. The surface

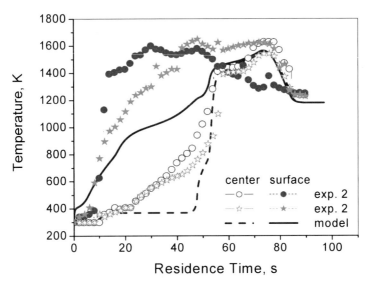

Figure 5.14 Temperature profiles of a near-spherical wet particle during combustion in air: $d_p = 11$ mm, $AR = 1.0$, $MC = 40$ wt%, $T_w = 1276$ K, $T_g = 1050$ K

contact is lost as the particle shrinks and the bead becomes exposed to the surrounding flame. The measured particle center temperatures appear to disagree with model predictions, though the disagreement arises primarily from thermocouple wire conduction. Both experimental data and model predictions show that during the char burning stage the particle temperature increases to a peak value and then declines dramatically. This supports theoretical descriptions of large-particle combustion mechanisms. Oxidizer diffusion rates primarily control combustion rates in char consumption, which proceeds largely with constant density and shrinking particle diameter. The char particle oxidation front will finally reach the center of the particle as particle size gets smaller with ash built up in the outer layer of the particle. The pseudo-steady-state combustion rate/temperature of the particle first increases then decreases with size due to changes in the relative importance of radiation losses, convection, and diffusion. Once the char is completely consumed the particle (ash) cools rapidly to near the convective gas temperature, depending on the radiative environment. The mass loss curves as functions of time are shown in Figure 5.15.

For a low moisture content (6 wt%), near-spherical particle ($d_p = 9.5$ mm, $AR = 1.0$), the flame temperatures are measured with both thermocouple and camera pyrometry. A type B thermocouple mounted near the particle surface provides some measurements of the flame temperature surrounding the particle. The upper limit of a type B thermocouple is about 2100 K, and the thermocouple data above this value is not accurate, as shown in Figure 5.16. The flame temperature was also interpreted by the imaging pyrometer with gray body emission assumption, where the results are combinations of flame and particle surface radiations. Both

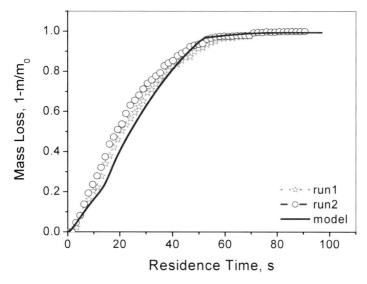

Figure 5.15 Mass loss of a near-spherical wet particle during combustion in air: $d_p = 11$ mm, $AR = 1.0$, $MC = 40$ wt%, $T_w = 1276$ K, $T_g = 1050$ K

thermocouple and pyrometry data are compared with model predictions in Figure 5.16, where the flame receded away from the thermocouple after devolatilization. The thermocouple measurements fluctuate due to the turbulence and two-dimensional effects caused by the bulk gas convection, which is not captured in this one-dimensional model. In the camera pyrometry measurements, soot was assumed as gray body emitter. The camera pyrometry measurements can be improved if spectral-dependent emissivity is applied in the calculation. The model prediction of the flame indicates the transition of combustion from devolatilization stage to char burning stage, appearing in Figure 5.16. Results show that model predictions generally agree with both the camera-measured data and thermocouple data, and the difference is within measurement uncertainty.

For a low moisture content (6 wt%), near-spherical particle ($d_p = 9.5$ mm, $AR = 1.0$), the flame temperatures are measured with both thermocouple and the camera pyrometry. A type B thermocouple mounted near the particle surface provides some measurements of the flame temperature surrounding the particle. The upper limit of a type B thermocouple is about 2100 K, and the thermocouple data above this value is not accurate, as shown in Figure 5.16. The flame temperature was also interpreted by the imaging pyrometer with gray body emission assumption, where the results are combinations of flame and particle surface radiations. Both thermocouple and pyrometry data are compared with model predictions in Figure 5.16, where the flame receded away from the thermocouple after devolatilization. The thermocouple measurements fluctuate due to the turbulence and two-dimensional effects caused by the bulk gas convection, which is not captured in this one-dimensional model. In the camera pyrometry measurements, soot was

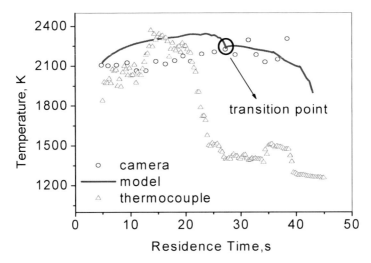

Figure 5.16 Flame temperature comparison during a near-spherical particle combustion in air: $d_p = 9.5$ mm, $AR = 1.0$, $MC = 6$ wt%, $T_w = 1273$ K, $T_g = 1050$ K

assumed as gray body emitter. The camera pyrometry measurements can be improved if spectral-dependent emissivity is applied in the calculation. The model prediction of the flame indicates the transition of combustion from devolatilization stage to char burning stage, appearing in Figure 5.16. Results show that model predictions generally agree with both the camera-measured data and thermocouple data, and the difference is within measurements uncertainty.

With the evaluated particle model, a series of predictions of different levels of complexity are compared with some sets of experimental data collected in both the entrained-flow reactor and the single-particle reactor, starting with an isothermal, spherical particle model, which is appropriate to pulverized coal particle.

5.8.4 Particle Temperature Measurements

Particle surface temperature data were collected for both sawdust particles in the long entrained-flow reactor and poplar particles in the single-particle reactor during pyrolysis and combustion.

5.8.4.1 Sawdust Particle Surface Temperature in the Entrained-flow Reactor

The particle surface temperatures were measured with the camera pyrometry installed around the entrained-flow reactor through optical accesses. Particle traveling speed in the reactor was determined with the imaging system. Figure 5.17 illustrates the sawdust particle surface temperature distribution when the particle was heated up in the reactor before a flame was formed around. A relatively poor image was obtained since the particle traveled at an average speed of about 3.0 m/s and the particle surface temperature was not high enough to use a very short exposure time. The average particle surface temperature was 1173 K, as indicated in the temperature map.

Once the particle starts to burn in the reactor, the particle itself will be surrounded by a flame. The flame may not completely block the radiation from the char particle surface, so both the flame and the particle surface contribute the signal received by the pyrometer sensor. The measured temperature will be a value between the char surface temperature and the flame/soot temperature. The measurement accuracy depends on the relative distance between soot cloud and the particle surface (which one is closer to the focus point of the imaging pyrometer), char surface temperature, flame temperature, soot absorption, *etc.* The temperature distributions of two burning particles in the entrained flow reactor appear below.

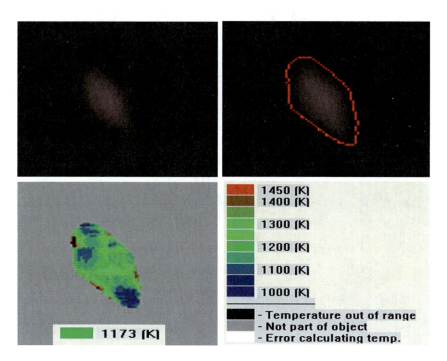

Figure 5.17 Particle surface temperature distribution during pyrolysis in nitrogen in entrained-flow reactor: $d_{p,eq} = 0.3$ mm, $MC = 6$ wt%, $T_w = 1312$ K, $T_g = 1124$ K

5.8.4.2 Poplar Particle Surface and Flame Temperature in Single-particle Reactor

Imaging pyrometers measure particle surface temperature and flame temperature during devolatilization and char burning through the optical accesses ports in the single-particle reactor wall. Thermocouples provide additional measurements of some of these data in many experiments. For the poplar pyrolysis, the algorithm determined both temperature and particle surface emissivity including reactor wall reflection correction, resulting in high values (up to 0.99) for charred surfaces.

The spatially and temporally resolved imaging pyrometer results indicate that combustion proceeds with non-uniform particle surface temperature. Specifically, the particle surfaces exposed to the most intense radiation (bottom) and, during oxidation, those at the leading edge in the induced convective flow (also the bottom) generally heat faster and to higher temperatures than the remaining particle surfaces.

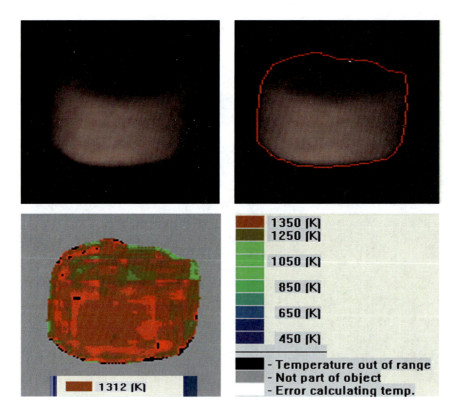

Figure 5.18 Poplar particle surface temperature during pyrolysis in nitrogen in a single-particle reactor: $d_p = 9.5$ mm, $AR = 1.0$, $MC = 6$ wt%, $T_w = 1373$ K, $T_g = 1050$ K

Three-dimensional rendering of these surface temperatures provides uniquely detailed data regarding combustion processes. To generate such renderings, the particle surface temperature distribution is mapped onto two-dimensional images, as illustrated in Figure 5.18. This figure illustrates data from a single residence time and for a near-spherical poplar particle suspended in the single-particle reactor with an average particle surface temperature of 1312 K, approaching the end of particle devolatilization.

Particle surface and flame temperature distributions have also been measured and mapped to 2D images for burning char particles, illustrated in Figures 5.19 and 5.20.

These data also indicate the flame temperature distribution when volatiles burn next to the particle during devolatilization. The average temperature of the flame is about 2200 K. In Figure 5.19, both the images of the char particle and the flame next to the particle appear in one frame, and the flame temperature and particle surface temperature are calculated and mapped simultaneously. Obviously, the average temperature of the flame zone is much higher than that in the particle surface zone, *i.e.*, during devolatilization the particle surface remains at a lower temperature than the surrounding flame. A char particle surface temperature map for a burning char particle appears in Figure 5.20.

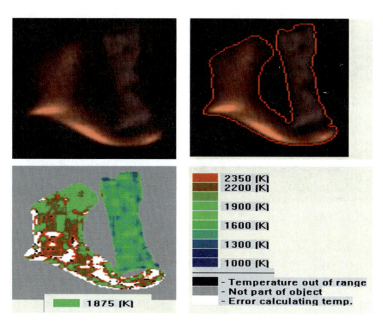

Figure 5.19 Char particle surface temperature and flame temperature map during particle devolatilization process: $d_p = 9.5$ mm, $AR = 4.0$, $MC = 6$ wt%, $T_w = 1273$ K, $T_g = 1050$ K

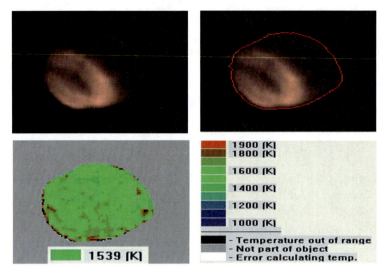

Figure 5.20 Char particle surface temperature map during char burning in air: $d_p = 9.5$ mm, $AR = 1.0$, $MC = 6$ wt%, $T_w = 1273$ K, $T_g = 1050$ K

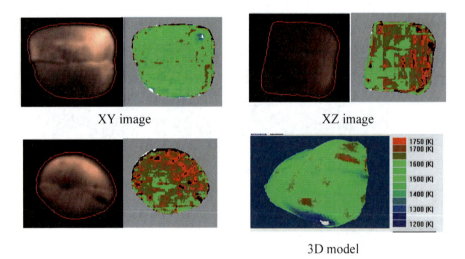

Figure 5.21 3D particle surface temperature map of a burning char particle: $d_p = 9.5$ mm, $AR = 1.0$, $MC = 6$ wt%, $T_w = 1273$ K, $T_g = 1050$ K

With the three camera pyrometers installed around the reactor, three images of a burning char particle were taken simultaneously from three orthogonal directions. The particle surface temperature from each angle was calculated individually for each image. The 3D particle shape, reconstructed with the three images as previously described, combines with the particle surface temperature distribution data to provide spatially and temporally resolved 3D particle data, as shown in Figure 5.21.

The time-resolved model predictions, thermocouple-measured and camera-measured temperature data, and the 3D particle surface temperature distribution data indicate that this transient, 1D particle combustion model captures the major particle combustion characteristics, but a 2D or 3D particle model can potentially improve the predictions for large particles with irregular shapes.

5.9 Conclusions

Experimental data from single particle reactors, including some first-ever experiments, include simultaneous and temporal data describing center temperature, spatially resolved surface temperature, size, shape, images from multiple directions, and mass. These simultaneous data reveal biomass reaction mechanisms and validate computer models. These mechanisms include, but are not limited to, particle drying, swelling, devolatilization, heat up, oxidation, and ash formation.

Particle drying proceeds by a modified Raoult's Law expression, where the mole fraction of water is replaced by a mass fraction expression. Mole fractions are difficult to define when dealing with biomass and similar natural components. Model predictions agreed with measurements within their uncertainty limits.

Devolatilization parameters yielded accurate amounts and rates of char and gas formation. This accounts for the largest and most rapid mass loss from biomass.

Char burnout predictions agree with data. Most chars burn at or near diffusion-limited rates; therefore these data provide little validation of char burnout kinetic parameters but do provide accurate predictions of burnout times and temperatures. They are representative of commercially significant biomass behavior.

An expression for the film thickness for a one-dimensional flame layer both predicts the presence of a flame and accounts for species heat and mass transfer in inert environments. This expression, suitable for the type of transient and one-dimensional model developed used in this document, does not account for the experimentally observed multidimensional flame structures around burning droplets/particles in suspension. Nevertheless, it appears to provide reasonably accurate estimates of flame effects on particle behavior.

Biomass particle size and shape have profound effects on overall particle combustion rates that the isothermal particle approaches commonly used for coal combustion cannot capture. Experimental evidence and modeling results indicate large internal temperature gradients, sometimes greater than $600°C/mm$, form during reaction, typically during drying/devolatilization. Experimental data and modeling evidence show large internal temperature and reaction/composition profiles in biomass particles of relevance to commercial systems. These profiles include complex dependencies on particle reaction rates and transport properties and change overall conversion times by factors of three or more for particles of relevance to commercial biomass applications. Inhomogeneous biomass fuels, such as straws, generate particles that require separate model descriptions for different sections of the fuel, *i.e.*, knees *vs* stalks in straw.

Models illustrated here capture most of these complexities. Nevertheless, the data indicate significantly more complex behaviors than can be captured in even the relatively sophisticated models used in this analysis. Specifically, large and potentially process-influencing temperature, reaction rate, and composition gradients along particle surfaces appear clearly in some experimental measurements but are not within the scope of models used here. The general predictive accuracy of the model suggests these effects may have limited practical implications.

References

1. Robinson AL, Rhodes J, Keith DW (2003) Assessment of potential carbon dioxide reductions due to biomass-coal cofiring in the United States, submitted to Environmental Science and Technology
2. Sami M, Annamalai K, Wooldridge M (2001) Cofiring of coal and biomass fuel blends. Prog Energ Combust 27:171–214
3. Hughes E (2000) Biomass cofiring: economics, policy and opportunities. Biomass Bioenerg 19:457–465
4. Baxter L (2005) Biomass-coal co-combustion: pportunity for affordable renewable energy. Fuel 84:1295–1302
5. Baxter L, Koppejan J (2004) Co-combustion of biomass and coal. Euroheat Power (English Edition) 34–39
6. Saxena SC (1990) Devolatilization and combustion characteristics of coal particles. Prog Energ Combust 16:55–94
7. Gavalas GR, Wilks KA (1988) Intraparticle mass transfer in coal pyrolysis. AICHE J 26: 201–211
8. Wildegger-Gaissmaier AE, Agarwal PK (1990) Drying and devolatilization of large coal particles under combustion conditions. Fuel 69:44–51
9. Russel WBS, PA, Greene MI (1979) AIChE Journal 25
10. Baxter LL (1993) Biomass Bioenerg 4:85–102
11. Tillman DA (2000) Biomass cofiring: the technology, the experience, the combustion consequences. Biomass Bioenerg 19:365–384
12. Kanury AM (1994) Combustion characteristics of biomass fuels. Combust Sci Technol 97:469–491
13. Di Blasi C (1996) Heat, momentum and mass transport through a shrinking biomass particle exposed to thermal radiation. Chem Eng Sci 51:1121–1132
14. Di Blasi C (1997) Influences of physical properties on biomass devolatilization characteristics. Fuel 76:957–964
15. Gronli MG, Melaaen MC (2000) Mathematical model for wood pyrolysis – comparison of experimental measurements with model predictions. Energ Fuel 14:791–800
16. Gera D, Mathur M, Freeman M, O'Dowd W (2001) Moisture and char reactivity modeling in pulverized coal combustors. Combust Sci Technol 172:35–69
17. Koufopanos C, Papayannakos N (1991) Modeling of pyrolysis of biomass particles: studies on kinetics, thermal and heat transfer effects. Can J Chem Eng 69:907–915
18. Williams A, Pourkashanian M, Jones JM (2001) Combustion of pulverized coal and biomass. Prog Energ Combust 27:587–610
19. Roberts W, Lu H, Ip L-T, Baxter LL (2009) Black liquor particle and droplet reactions: experimental and model results. 45th Anniversary International Recovery Boiler Conference, Lahti, Finland, pp 57–100
20. Baxter LL (1993) Ash deposition during biomass and coal combustion: a mechanistic approach. Biomass Bioenerg 4:85–102

21. Miles TR, Miles TR Jr, Baxter LL, Bryers RW, Jenkins BM, Oden LL (1996) Boiler deposits from firing biomass fuels. Biomass Bioenerg 10:125–138
22. Jenkins BM, Kayhanian M, Baxter LL, Salour D (1997) Combustion of residual biosolids from a high solids anaerobic digestion aerobic composting process. Biomass Bioenerg 12:367–381
23. Baxter LL (1998) Influence of ash deposit chemistry and structure on physical and transport properties. Fuel Process Technol 56:81–88
24. Baxter LL, Miles TR, Jenkins BM, Milne T, Dayton D, Bryers RW, Oden LL (1998) The behavior of inorganic material in biomass-fired power boilers: field and laboratory experiences. Fuel Process Technol 54:47–78
25. Jenkins BM, Baxter LL, Miles TR Jr, Miles TR (1998) Combustion properties of biomass. Fuel Process Technol 54:17–46
26. Robinson A, Junker H, Buckley SG, Sclippa G, Baxter LL (1998) Interaction between coal and biomass when cofiring. Twenty-Seventh Symposium (International) on Combustion/The Combustion Institute 1351–1359
27. Kaufmann H, Nussbaumer T, Baxter L, Yang N (2000) Deposit formation on a single cylinder during combustion of herbaceous biomass. Fuel 79:141–151
28. Nielsen HP, Baxter LL, Sclippab G, Morey C, Frandsen FJ, Dam-Johansen K (2000) Deposition of potassium salts on heat transfer surfaces in straw-fired boilers: a pilot-scale study. Fuel 79:131–139
29. Nielsen HP, Frandsen FJ, Dam-Johansen K, Baxter LL (2000) Implications of chlorine-associated corrosion on the operation of biomass-fired boilers. Prog Energ Combust 26: 283–298
30. Robinson AL, Junker H, Baxter LL (2002) Pilot-scale investigation of the influence of coal-biomass cofiring on ash deposition. Energ Fuel 16:343–355
31. Turn SQ, Kinoshita CA, Jakeway LA, Jenkins BM, Baxter LL, Wu BC, Blevins LG (2003) Fuel characteristics of processed, high-fiber sugarcane. Fuel Process Technol 81:35–55
32. Bharadwaj A, Baxter LL, Robinson AL (2004) Effects of intraparticle heat and mass transfer on biomass devolatilization: experimental results and model predictions. Energ Fuel 18:1021–1031
33. Kær SK, Rosendahl LA, Baxter LL (2006) Towards a CFD-based mechanistic deposit formation model for straw-fired boilers. Fuel 85:833–848
34. Lokare SS, Dunaway JD, Moulton D, Rogers D, Tree DR, Baxter LL (2006) Investigation of ash deposition rates for a suite of biomass fuels and fuel blends. Energ Fuel 20:1008–1014
35. Damstedt B, Pederson JM, Hansen D, Knighton T, Jones J, Christensen C, Baxter L, Tree D (2007) Biomass cofiring impacts on flame structure and emissions. P Combust Inst 31: 2813–2820
36. Wu C, Tree D, Baxter L (2007) Reactivity of NH_3 and HCN during low-grade fuel combustion in a swirling flow burner. P Combust Inst 31:2787–2794
37. Lu H, Robert W, Peirce G, Ripa B, Baxter LL (2008) Comprehensive study of biomass particle combustion. Energ Fuel 22:2826–2839
38. Ip L-T (2005) Comprehensive black liquor droplet combustion studies. PhD thesis. Department of Chemical Engineering, Brigham Young University
39. Ip L-T, Baxter LL, Mackrory AJ, Tree DR (2008) Surface temperature and time-dependent measurements of black liquor droplet combustion. AICHE J 54:1926–1931
40. Lu H, Ip L-T, Mackrory A, Werrett L, Scott J, Tree D, Baxter L (2009) Particle-surface temperature measurements with multicolor band pyrometry. AICHE J 55:243–255
41. Ip LT, Baxter LL, Mackrory AJ, Tree DR (2008) Surface temperature and time-dependent measurements of black liquor droplet combustion. AICHE J 54:1926–1931
42. Lu H, Ip L-T, Mackrory A, Werrett L, Scott J, Tree D, Baxter L (2009) Particle surface temperature measurements with multicolor band pyrometry. AICHE J 55:243–255

Chapter 6
Fluidized Bed Combustion of Solid Biomass for Electricity and/or Heat Generation

Panagiotis Grammelis, Emmanouil Karampinis and Aristeidis Nikolopoulos

Abstract Fluidised bed combustion (FBC) technology was developed in the 1970s in order to exploit the energy potential of high-sulphur coals in an environmentally acceptable way. The FBC technology was soon expanded for biomass and other low-grade fuels, which have typically large variations in fuel properties. The benefit of the FBC is the large amount of bed material compared with the mass of the fuel (98 *vs* 2%) and, thus, the large heat capacity of the bed material that stabilises the energy output caused by variations in fuel properties. Moreover, by selecting reagents as bed material and controlling the bed temperature, the emissions of pollutants can be controlled. In the last two decades, rapid progress has been achieved in the application of FBC technology to power plants up to intermediate capacities, caused by the increasing demands for fuel flexibility, stringent emission control requirements, stable plant operation and availability. Especially concerning the fuel range; there is a definite trend to widen the range of biomass fuels and waste fractions. The aim of this chapter is to review critically the technical requirements of biomass and/or waste combustion in FBCs, the operational problems, the needs for emissions control and the ash handling issues.

6.1 Introduction

Fluidised bed combustion was developed in the 1970s aiming to utilize high-sulphur coals for energy production, while remaining within the acceptable limits

Emmanouil Karampinis (✉)
Centre for Research & Technology Hellas/
Institute for Solid Fuels Technology & Applications (CERTH/ISFTA),
Mesogeion Ave. 357-359, GR-15231 Halandri, Athens, Greece
Tel: +30 210 6501593, Fax: +30 210 6501598
e-mail: karampinis@certh.gr

of the environmental legislation and efficiency. The central idea of fluidized bed technology is the burning of fuel in an air-suspended mass (or bed) of inert particles. Due to the large amount of bed material compared to the fuel mass (98 *vs* 2%), the heat capacity of the inert particles is large and capable of stabilizing the energy output, despite variations in the heating value of the fuel. Moreover, FBC offers the potential to limit the environmental impact of solid fuel combustion, by controlling SO_x and NO_x emissions. The former is achieved by the correct choice of bed material additives, such as dolomite or limestone, which absorbs the emitted sulphur, while the formation of the latter is kept to a minimum by setting the combustion temperature at comparatively low levels (800–900°C) [1, 2].

In fluidized bed combustion, the solid bed material is normally made of inert materials, such as silica sand and/or ash, with the possible addition of a sorbent, such as limestone. Initially, the solid particles are located at the bottom part of the boiler, over a plate with air distribution nozzles. As increasing quantities of primary air are fed through the nozzles, the drag forces on the particles counteract the gravitational ones, till the minimum fluidization velocity is reached. A further increase in air velocity results in a fluidized bed of materials. Depending on the air velocity, two major types of FBC are commonly utilized. A Bubbling Fluidized Bed (BFB) is characterized by air velocities in the range of 1.0–3.0 m/s. The overall impression is of a violent, non-uniform, bubbling fluid, which includes bubbles of air amidst the bed material. The most important characteristic of a BFB is that particles do not leave the bed area, unless their particle size is greatly reduced. On the other hand, Circulating Fluidized Beds (CFB) exhibit higher air velocities, in the range of 3.0–6.0 m/s. As a result, part of the bed material is constantly leaving the bed. In order to keep the bed material in a constant fuel ratio, the outgoing particles are collected by cyclone separators and recirculated into the furnace. In CFBs there is no longer a clear distinction between the dense bed zone and the dilute upper zone, as is the case in BFBs; the solid density is decreasing proportionally to the furnace height.

Both FBC concepts are operated at low temperatures, typically ranging from 800 to 900°C. Temperature control is achieved by internal heat exchanger surfaces, flue gas recirculation, water injections and/or sub-stoichiometric bed operation [1]. Secondary air can also be employed and is introduced above the bed area through inlets located throughout the boiler width. A simple schematic diagram of a common CFB and BFB configuration is presented in Figure 6.1, while Table 6.1 summarizes some of the differences in those two concepts.

Overall, FBC present a number of advantages over conventional firing systems, such as [3–5] (a) the ability to utilize fuels of varying quantity in terms of size, shape, moisture, ash and heating value; the range remains wide once the furnace is built, (b) a high heat transfer rate between bed and heat exchangers, (c) stable, low-temperature combustion conditions, (d) good control of the facility (FBCs do not require a hot-restart even in cases where the fuel supply is interrupted for several minutes), (e) high combustion efficiency; (f) low NO_x (practically, no thermal NO_x is produced) and SO_x formation.

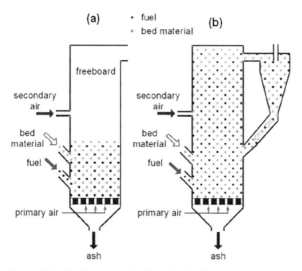

Figure 6.1 (a) Circulating fluidized bed boiler, and (b) Bubbling fluidized bed boiler [1]

However, FBC systems also possess several disadvantages, some aggravated by the low quality of the employed fuels. These include [6, 7] (a) the need for highly efficient gas-solid separation systems due to high dust loads, (b) high erosion rate of internal surfaces due to high solid velocities (especially in the case of CFB), (c) agglomeration of bed particles, which may lead to plant shutdowns, (d) inherent pressure drop losses in the riser which decrease the overall efficiency of the power plant.

In the following sections, issues related to the operation of CFB utilizing solid biomass and/or waste fuels will be discussed. Those issues can generally be categorized as related to fuel supply and quality (*e.g.* variations in fuel properties, pretreatment), operation (*e.g.* agglomeration, corrosion) and finally environmental aspects (*e.g.* gas emissions, ash, trace elements).

Table 6.1 Comparison of BFB and CFB technologies

	Bubbling Fluidized Beds	Circulating Fluidized Beds
Air velocity (m/s)	1.0–3.0	3.0–6.0
Bed material size (mm)	<0.5–1.0	<0.2–0.4
Fuel particle diameter (mm)	<80	<40
Excess air (%)	20–30	10–20
Bed Temperature (°C)	650–850	750–900
Capacity (MW$_{th}$)	>20	>30
Tar in flue gas	Moderate	Low
Dust load in flue gas	Very high	Very high (higher than BFB)
Combustion efficiency (%)	90–96	95–99.5
Heat transfer rate (MW/m^2)	0.5–1.5	3.0–4.5
Specific investment[a]	Lower	Increased[a]

[a]Due to increased boiler size and possible requirements for fuel pre-treatment

6.2 Biomass/Waste Fuel Properties

The term biomass covers a wide range of biological materials. Moreover, it is also applied to the biogenic content of waste streams. Regarding energy exploitation, several distinct biomass and/or waste types are commonly utilized, for example agricultural, forestry and industrial residues, municipal waste and energy crops. For a more detailed discussion on biomass and waste fuel categorization and characterization methods, the reader is referred to Chapters 2 and 4 of this work.

Typical analyses results for several biomass fuels are reported in Table 6.2 [8–10], while ash analyses, compared to solid fossil fuels, are presented in Table 6.3. Although chemical and physical properties of the biomass/waste fuel differ on source and pre-treatment, most fuel types share some important common characteristics which influence their combustion behaviour. These are the high volatile content and the presence of oxygen, which influences the heating value range. Other important properties affecting the combustion behaviour of biomass/waste fuels are moisture content, chlorine concentration and ash characteristics. Issues related to physical properties, such as density and particle size, will be discussed in Section 6.3.

6.2.1 Volatile Content

Biomass volatile content is typically in the range of 60–80 wt%. As a result, most of the weight loss exhibited during combustion occurs during the devolatilization phase. Moreover, most of the heat contribution of biomass combustion is due to volatile matter [11]. Volatile release rate is rapid, resulting in high reactivities for biomass fuels. Typical products of the devolatilization phase include light hydrocarbons, carbon monoxide, carbon dioxide, moisture, hydrogen, hydrogen cyanide, ammonia and tars. Yields and rate of release are dependent on fuel characteristics, such as lignin content, as well as on temperature and heating rate.

Volatile matter is mostly burned to the freeboard zone above the bed. In order to guarantee high combustion efficiencies, as well as to reduce unburnt pollutant emissions, such as CO and PAH, high residence times need to be guaranteed. Many CFB boilers are designed for high-volatile content coals, and therefore such problems do not manifest during biomass firing or co-firing. Moreover, biomass and coal properties may combine to produce a better combustion behaviour compared to the individual fuels [12].

The high volatile content of biomass and the low degree of heat release in the lower bed provided a benefit for BFB applications, since it makes staged combustion particularly suitable [3]. Staged combustion can achieve a reduction of heat extraction surfaces via tubes immersed in the bed, which erode frequently and cause availability problems.

Table 6.2 Analysis of typical biomass fuels [8–10]

Fuel	Proximate analysis (wt%, as received)				Ultimate analysis (wt% db)					H.H.V.[d] (MJ/Kg)
	Moisture	V.M.[a]	F.C.[b]	Ash	C	H	N	S	O[c]	
Olive kernel	18.1	60.8	18.5	2.6	51.6	8.5	1.08	0	38.82	29.86
Almond shell	9.7	66.9	20	3.36	49.6	6.4	0.6	0	43.4	26.21
Forest residue	21.3	63.7	13.7	1.2	51	7.9	0.3	0	40.8	28.82
MDF	6.5	74.4	17.9	1.1	47	6.4	4.6	0	42	25.29
Saw dust	14.3	73.3	11.96	0.44	51	7.9	0.3	0	40.8	28.82
Waste wood	5.63	69.53	16.08	8.76	45.99	5.67	3.82	0.05	35.44	23.91
Willow	11.3	71.47	16.1	1.14	44.7	5.7	0.2	0.03	48.21	16.97
Demolition wood	8.6	70.69	18.8	1.92	44.5	5.6	1.1	0.09	46.52	17.87
Straw	8.0	71.7	15.0	5.3	41.37	6.03	0.47	45.23	0.19	16.5
MBM	1.35	79.66	8.61	10.38	55.67	8.03	7.15	0.05	29.1	30.64
MBM Demineralized	0.52	79.83	19.36	0.29	52.87	7.8	7.34	0.79	31.2	29.42
Cynara cardunculus[e]	8.2	70.0	14.6	7.2	40.6	5.5	0.9	0.1	45.0	13.7
Cynara cardunculus[f]	7.9	71.8	13.4	6.9	46.7	4.8	0.7	0.1	47.7	16.3

[a]Volatile matter
[b]Fixed carbon
[c]By subtraction
[d]Higher heating value
[e]Excluding seed
[f]Including seed

Table 6.3 Ash analysis of typical biomass [8.10]

Fuel	Ash (wt%)										
	SiO$_2$	Al$_2$O$_3$	Fe$_2$O$_3$	MnO	TiO$_2$	CaO	K$_2$O	MgO	P$_2$O$_5$	Na$_2$O	SO$_3$
Olive kernel	29.09	5.39	1.84	0.05	0.11	37.23	10.48	2.26	3.46	1.69	3.82
Cotton residues	28.07	5.43	1.06	0.04	0.09	18.5	13.38	11.39	7.09	4.88	8.83
Forest wood	17.25	5.56	8.44	–	–	32.66	16.32	7.89	–	2.02	16
Waste wood	31.83	8.32	2.76	0.29	3.85	33.47	1.46	6.5	0.78	2.48	7.6
Paper pulp	25.85	14.1	0.71	0.01	0.23	53.22	0.34	5.08	0.07	0.01	0.39
Wood pellets	27.43	10.48	3.17	0.3	2.15	27.54	3.6	5.13	4.44	3.61	12.25
Greek lignite	20.91	9.45	5.96	0.03	0.4	47.29	0.2	3.34	0.19	0.17	10.04
Hard coal	43.37	23.95	10.04	0.13	1.23	5.95	3.05	4.26	0.81	0.99	5.92

6.2.2 Heating Value

Heating value is generally correlated to the presence of C, H and O, with an increase attributed to higher percentages of the first two elements, while a decrease is observed with higher decrees of oxidation. Due to the increased oxygen concentration, the typical range of heating values for biomass fuels is lower than that of bituminous coals; however, biomass exhibits increased heating values compared to several lignite varieties.

The heating value in biomass and waste fuels is also correlated to the presence of cellulose, hemicellulose and lignin. Of those three components, lignin possesses the highest heating values, while hemicellulose the lowest, due to its high oxidation degree [13].

The effect of the heating value of the fuel on the combustion system is manifested in the plant design and the utilization level. For co-firing applications where the substitution range does not exceed 10% of thermal share, it is assumed that the effect of the lower quality fuel does not have a significant effect on the combustion performance.

6.2.3 Moisture

The biomass moisture content varies widely, depending mostly on the biomass type and the pre-treatment method employed. Typical values are 25–60 wt%. For agricultural residues and energy crops, weather conditions and harvesting season have an important effect on moisture content. Low moisture (< 10 wt%) is found in dry-wood processing residues or biofuels in pellet form.

High moisture content has an adverse effect on biomass handling and storage, contributing to dry-matter loss through decomposition and feeding blockages. Another adverse effect of high moisture is the delay of ignition, the higher time required for drying, which delays the devolatilization and char combustion stages, and a reduction in the adiabatic temperature, which decreases the burnout of the volatiles and char [11]. Moreover, the heating value of the biomass fuel is decreased due to the energy consumption for evaporation. Therefore, in dedicated biomass installations, a supporting fuel, such as natural gas, may be required. Moreover, a higher volume of flue gas is produced due to increased fuel utilization. Overall, high moisture content requires a larger combustion chamber and an increase in the dimensions and capacity of the flue gas cleaning systems.

6.2.4 Chlorine Content

The most important differentiation of biomass and coal fuel is their chlorine content. Chlorine, which is typically not found in coals, is present in biomass in percentages ranging from less than 0.1% to more than almost 2 wt% (dry fuel) [14]. Approximately, each 0.1 wt% of chlorine in the fuel corresponds to approximately 100 ppmv chlorine in the gas phase.

During biomass combustion, chlorine is almost completely vaporized, forming HCl, Cl_2 and alkali chlorides. The importance of chlorine stems from the impact of the alkali chlorides on high temperature corrosion of superheater tubes and on the influence of HCl in the creation of PCDD/F. Moreover, chlorine facilitates the transport of volatile heavy metals from the fuel ash to aerosol particles.

6.2.5 Ash Characteristics

As is the case with other properties, the ash content of biomass fuels varies significantly among different varieties. For example, ash content is less than 1% in many wood ash types and low values are reported for many agricultural residues. On the other hand, high ash content is typically found in sewage sludge and rice hulks [10]. Generally, though, the ash content of biomass fuels can be handled by the flue gas cleaning systems.

The most important issue concerning biomass ash is not quantity but quality. Typically, biomass ash contains elements not found in coal ash, such as alkali metals and silica. The presence of alkalies and silica, which is quite high for herbaceous fuels such as agricultural residues, lowers the ash melting temperature [10, 11], while calcium and magnesium increase it. A significant portion of alkalis is released in the gas phase, especially in the presence of chlorine. However, the amount remaining in the solid phase is enough to form low-temperature melting compounds with silica, which leads to agglomeration issues and unit shutdowns due to low quality fluidization.

Other issues related to biomass ash include its utilization potential, for example the effect of heavy metals in ash utilization in cement kilns [15], and the presence of heavy elements in certain kinds of biomass, such as demolition wood or sewage sludge. Moreover, a secondary repercussion is the possible deactivation of the catalysts in the SCR (selective catalytic reduction) systems [16] by the alkaline earth metals in biomass ashes. For more information, the reader is referred to Chapter 10.

6.3 Operational Issues of Biomass-fired FBCs

The examination of the physical and chemical characteristics of biomass fuels clearly illustrates that they differ in several important aspects from solid fossil fuels, such as hard coal and lignite that are typically employed in combustion systems. These differences have a significant impact on several aspects of the operation of an FBC and have to be taken into account during the design of a new unit or the retrofitting of an older one. In this section, three of the most important operational challenges that a biomass-fired FBC unit faces are examined: handling and feeding issues, deposition/corrosion and agglomeration.

6.3.1 Biomass Handling and Feeding

The design of the biomass handling system is one of the most critical areas of a biomass FBC system. Many biomass plants experience significant problems during their initial operation, such as fuel pile odours, heating and decomposition, equipment wear, bottlenecks in the feed system and fluctuations in moisture content [1]. Several pre-treatment processes can be applied in order to homogenize or improve biomass combustion qualities. These include shredding, chipping and grinding for size reduction and drying for moisture removal.

Densification is one of the most important pre-treatment steps, since low density (close to 100 kg/m^3) is a typical characteristic of biomass fuels and has extremely important economic and technical repercussions. Apart from economic concerns, such as extensive transportation and large storage area costs, low density fuels present feeding issues, process control difficulties and fuel entrainment problems. Baling, briquetting and pelletization are the most commonly used methods for biomass densification [2]. Baling is commonly applied for agricultural biomass in order to increase energy density and ease of handling. The bulk density of bales depends on both the machine and the biomass type; typical values are in the rage of 100–150 kg (dry matter)/m^3 [1]. Briquetting and pelletization can achieve more significant density increases, in the range of 450–650 kg/m^3 [2]. They are typically applied for fine wood particles, such as sawdust, although pelletization of agricultural biomass is also becoming increasingly common. Briquetting is a simpler technology and, in contrast to pelletization, does not require previous chopping and milling steps. Pelletization results in lower particle sizes, can achieve higher press outputs and is more tolerant of variations in the moisture content [2]. The energy cost of both these technologies is quite high, reaching up to 2.5% of the NCV for pelletization, which can be increased to 20% if a drying step is required [1].

Handling and storage options can be loosely classified based on the biomass type used. Two major classifications exist: harvested, mostly herbaceous biomass and non-harvested, mostly woody biomass.

Harvested fuels include long and slender biomass types, such as straw and grass. Such herbaceous fuels are typically pressed into bales and are transported in this form from the field to the fuel yard. Storage has to take into account the fact that exposed bales are subject to rainfalls, which include the moisture content and increase the rate of decomposition [1]. Piling is acceptable only for short-term storage, while for long-term storage flying roofs or indoor storage must be employed in order to ensure fuel quality. Bales are typically handled with wheeler loaders or crane systems and must pass through a bale shredder and a rotary cutter chopper in order to reach the acceptable particle size for the combustion system [3].

Non-harvested biomass fuels, such as wood, bark, prunings and residues of food processing industries, are typically granular in shape. Trucks, wagons or barges are common means of transport. Woody biomass types are usually delivered in chips, while food processing by-products (*e.g.* nuts, shells) come in bulk particles [3]. Piling is the most common method of storage, although care must be taken to avoid self-ignition issues, due to the temperature increases in the core of the pile. Bark is especially prone to such incidents [1]. For biomass fuels with small particle sizes, such as sawdust, outdoor storage is discouraged due to dust emissions [2].

Several feeding systems are used depending on the biomass properties and size, as well as on the combustion technology employed. These include pneumatic conveyors, screw feeders, moving hole feeders and ram feeders [1, 3]. Screw feeders, especially multiple ones, are amongst the most common for biomass sizes less than 50 mm, *e.g.* sawdust and pellets. However, their application is limited to small beds and may exhibit plugging problems when handling high moisture biomass [3]. Ram feeders are also employed for more sticky or fibrous materials [3].

BFBs use either the over-bed or under-bed feeding systems. Over-bed systems do not require very small particle sizes or pneumatic injection lines, thus keeping the overall fuel pre-treatment cost low. However, for typical coals their combustion efficiency is decreased due to the relatively high number of particles that escape the bed unburnt. For biomass fuels though, the reactivity is high and over-bed systems are used most frequently [3]. In CFBs, the feeding point is usually located in the lower, sub-stoichiometric part of the bed in order to ensure adequate resident times. Biomass is usually fed into the loop-seal, where it is partially de-volatilized before entering the bed in a well-mixed state [3].

6.3.2 Deposition/Corrosion

A common problem associated with solid fuels energy applications is the gradual reduction of the heat transfer rates due to the deposits accumulated on the heat exchanging equipment. The term slagging is used to characterize deposits formed on sections of the boiler exposed mainly to radiant heat, such as the furnace walls, while fouling refers to deposits formed on the convective pass, such as the heat exchanger tubes. Slagging and fouling increase reduce heat transfer and facilitate the initiation of corrosive reactions. Corrosion refers to the deterioration of intrinsic properties of the wall material is caused by complex reactions involving gas phase species, deposits or their interaction. Although slagging and fouling problems affect the overall efficiency and the availability of the equipment, soot-blowing (for superheater tubes) and cleaning during plant shut-downs can remove deposits. Corrosion, on the other hand, is permanent and severely affects the lifetime of the equipment.

Alkali compounds, sulphur and chlorine are the most important species involved in fouling/corrosion phenomena during biomass combustion [17–19]. An overview of the pathways for these chemical species is presented in Figure 6.2.

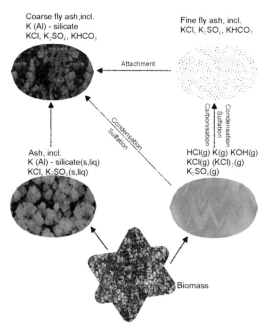

Coarse fly ash,incl.
K (Al) - silicate
KCl, K₂SO₄, KHCO₃

Fine fly ash, incl.
KCl, K₂SO₄, KHCO₃

Attachment

Ash, incl.
K (Al) - silicate(s,liq)
KCl, K₂SO₄(s,liq)

HCl(g) K(g) KOH(g)
KCl(g) (KCl)₂(g)
K₂SO₄(g)

Biomass

Figure 6.2 Alkali and chlorine interactions during biomass combustion [19]

Part of the alkali content of the biomass ash is released in the gas phase through the formation of chlorides or hydroxides. Chlorine in particular facilitates the vaporization of the alkali metals more strongly than the alkali concentration itself [2]. The rest of the alkalis remain in the fuel ash, in the form of silicates of sulphates, and ends up in the coarser fly ash particles, which are mainly constituted of refractory species, such as Ca, Mg and Si. On the other hand, gas phase alkali condenses in fine ash particles. A large part of these aerosols escape with the stack gas, while parts may be attached to the coarse ash particles or condense on the tubes. The alkali silicates in ash have low melting temperatures, sometimes less than 700°C [2]. Thus, molten ash particles build up on heat transfer surfaces as sticky deposits, which enhance deposition by the adherence of even coarse fly ash particles that would normally bounce off the surfaces.

Corrosion phenomena occur when the protective oxide layer that is formed on tube walls is attacked by chlorine or sulphur containing compounds. The sulphidation and chloridation of the tube surfaces results in the formation of an outer layer that does not have the protective properties of the oxidised one. Its defective structure means that it can be scaled off easily due to erosion and thus become subject to further corrosion [3]. Corrosion can take place either through gas phase reactions of compounds such as Cl_2 and $NaCl_{(g)}$ with the metal surface or through solid and molten phase reactions with sulphates and chlorides.

Sulfidation reactions are not common for biomass-fired FBCs due to the low sulphur content of the fuel. Chlorine corrosion is much more important and can occur either through chloride containing species in the gas phase or through chloride enriched deposits that accumulate on the tube surfaces according to the mechanisms described above. A detailed description of the chlorine based corrosion can be found in Chapter 5.

As indicated above, biomass fuels with high concentrations of alkali species and chlorine, such as straw and many agricultural residues, are expected to present severe ash deposition and corrosion problems at high or moderate combustion temperatures. The use of additives, such as bauxite, kaolinite and limestone, for the creation of high temperature melting alkali compounds has been suggested as a remedy by several studies [20–22]. However, in these cases, fuel bound chlorine is still released in the gas phase and gas phase corrosion reactions may still take place [2].

6.3.3 Agglomeration

As mentioned previously, part of the fuel ash in an FBC remains in the bed. It is this ash which is the main cause for the agglomeration problems experienced in FBCs, that is the adhesion of bed particles due to the melting of part of the fuel ash. Agglomeration has always been a major issue for fluidized bed combustion and has been studied for coal-fired applications for many years. The major characteristic of many biomass ashes, *e.g.* their low melting temperature, typically aggravates the problem.

Several factors affect the agglomeration phenomenon, such as alkali and alkali earth concentrations in the ash, temperature and fluidization temperature [23, 24]. For biomass fuels, the most important elements are potassium and sodium, which decrease the agglomeration time, and calcium and magnesium, which tend to increase it, subject to restrictions imposed by the sodium concentration and temperature [24]. The alkali oxides or salts remaining in the bed react with the quartz bed material, which is mostly SiO_2, according to the following reactions:

$$2SiO_2 + Na_2CO_3 \rightarrow Na_2O \cdot 2SiO_2 + CO_2 \qquad (6.1)$$

$$2SiO_2 + K_2CO_3 \rightarrow K_2O \cdot 2SiO_2 + CO_2 \qquad (6.2)$$

The mixtures formed are eutectic, with melting temperatures of 874°C and 764°C respectively, which are both lower than the individual components [25, 26] and the SiO_2 melting point, which is around 1,450°C [27]. Surveys indicate that the most likely alkali species transfer mechanism relies on collisions of sand with burning char particles [28]. Moreover, it was suggested that bed agglomerates start to form near burning char, where the local temperature is higher, enhancing the melting of the ash and the particle stickiness. This was verified by further experimental work, which reported that agglomerate particles are typically hollow, as a result of the char particle which initiated the whole process [29].

The creation of large particles due to the agglomeration mechanism disturbs the fluidization and gives rise to local hot spots. These hot spots in turn enhance the melting and sintering of particles and further aggravate the agglomeration issue. A continuation of the fuel feeding extends the agglomeration and may lead to defluidization of the bed and plant shutdowns [27].

The existence of agglomerate particles in the bed can be detected by the presence of temperature gradients and pressure drop fluctuations in the bed [28]. These are usually interpreted as the result of a reduction of uniform mixing conditions of the bed and a faster accumulation of ashes. Early detection may prevent further agglomeration in case corrective measures are applied. In other situations, agglomeration commences very early, requiring more drastic and pre-emptive measures. The main idea of these measures is to increase the melting point of sintering compounds.

The use of additives, such as kaolin, dolomite, limestone and alumina, has been proposed [30]. However, their use has been limited by efficiency issues and other subsequent problems [2]. Co-firing with coal or other, non-problematic biomass fuels has also been suggested as a potential remedy. The sulphur content of the coal in particular helps reduce the agglomerate formation [32]. Moreover, alkali metals from biomass interact with the clay mineral present in coals to form alkali-alumina silicates. In high sulphur conditions, alkali sulphates are formed, which have higher melting points [3]. The pre-treatment of fuel is also used to remove low-temperature melting compounds. For example, in Denmark, straw is left in the fields in winter and cut in late spring, thus allowing the rain water to leach the alkali away from the fuel [30].

The most attractive solution for plant operators is the use of alternative bed materials, due to its ease of use and relatively low costs [2]. Alternative materials considered include feldspar, dolomite, magnesite and alumina, which form eutectic mixtures but with high-melting temperatures [3]. Iron oxide, whether as an additive or as a constituent of the fuel ash, reduces the rate of agglomerate formation, since it reacts preferentially with the alkali compounds (represented as X), according to the reactions:

$$Fe_2O_3 + X_2O + \rightarrow X_2Fe_2O_4 \tag{6.3}$$

$$Fe_2O_3 + X_2CO_3 \rightarrow X_2Fe_2O_4 + CO_2 \tag{6.4}$$

The eutectic mixture has a melting temperature higher than $1135°C$ [25, 26], which is higher than the usual operating temperatures of FBCs. However, problems, such as high attrition and entrainment rates, chemical stability and windbox and air nozzle plugging issues, have been reported for these methods [2]. A new patented bed material, called "Agglostop", has also been reported as successful in dealing with the agglomeration issues in several plants handling fuels with high alkali concentrations [30, 31]. The ECOFLUID® and BioCOM® technologies have also been reported to deal with the agglomeration issue by utilizing staged combustion, which amounts to sub-stoichiometric combustion in the bed and the retention of the bed temperature at levels below the melting point of most ashes [33, 34, 37].

6.4 Environmental Aspects

The main environmental advantage of biomass is its status as a renewable fuel, with no or negligible impact on CO_2 emissions. However, as with all solid fuels, other environmental concerns over its combustion have been raised. The pollutants associated with biomass combustion can generally be categorized as unburnt pollutants or pollutants produced by the combustion [2]. The first category includes PAHs, dioxins, CO and char particles, which indicate a problem in the combustion efficiency. Therefore, methods for their reduction include all measures intending to increase the combustion efficiency, such as more efficient mixing of combustibles and air, temperature and residence time increases and staged combustion. The second category includes pollutants such as PM, oxides of nitrogen and sulphur, as well as acid gases and heavy metals, which are not a product of incomplete combustion. Their formation is a function of biomass properties and combustion conditions and can be controlled either by fuel pre-treatment or by careful modifications of stoichiometry or other combustion parameters. Moreover if all the aforementioned techniques fail to meet the imposed environmental limits, post-combustion flue gas cleaning (SCR, *etc.*) are adopted. In the following sections, the formation mechanisms for the most important pollutants in FBC conditions will be outlined and methods of prevention or reduction will be presented.

6.4.1 PAHs

Polycyclic aromatic hydrocarbons (PAHs) are organic chemical compounds that consist of fused aromatic rings. PAHs typically contain four to seven member rings, though five or six is the most common variety [35]. Not all PAHs exhibit the same toxicity, since their effect on humans varies depending on their structure; however, many are known or suspected for carcinogenic properties.

Two main mechanisms for PAH formation have been proposed:

1. Incomplete combustion: this mechanism is relevant for fuels whose structure is mainly aromatic. Unburned fractions of the fuel are emitted in the gas phase as PAHs.
2. Pyrolysis and pyrosynthesis: in fuel-rich regions of the flame, polymerization reactions are often favoured over oxidation. Such reactions include the cyclization of alkyl chains and radical condensation [36, 37].

The first mechanism is relevant for coal, which has high concentrations of polycyclic aromatic compounds. Biomass, on the other hand, contains less aromatics, thus making pyrolysis and pyrosynthesis the most relevant mechanism for PAH formation [2]. The PAH producing mechanisms are generally very complex and depend both on the combustion conditions and the fuel properties. Usually, the Hydrogen Abstraction Acetylene Addition Mechanism (HACA) is proposed as the most common mechanism for the creation of PAHs from small C_2 species. Larger PAHs are produced from more simple forms through condensation and cyclization. Typically, three-ring PAHs are found in the gas phase, while larger compounds are supported on soot and fly ash particles, or, in the case of the largest compounds, form particles of their own [2]. A simplified scheme for PAH formation, including the mechanisms for producing PCDDs and PCDFs (see next section), is presented in Figure 6.3 [38].

In FBC conditions, PAH formation is enhanced at temperatures higher than 850°C, since the synthesis reaction of PAH formation are endothermic. Moreover, the presence of metals, such as Fe and Cu, in the bed material may also act as a catalyst [39]. The presence of limestone can also have a negative effect on PAH formation, due to the endothermic decomposition of limestone and the subsequent perturbations of the thermal balance and the longer turnover of bed material. As a result, local fuel-rich spots, which produce higher amounts of PAHs, may be formed. Intensive mixing, which is an inherent characteristic of FBC, alleviates the avoidance of such spots. Nevertheless, in large scale installations (typically above 200MW$_e$) efficient mixing can be challenging.

The reduction of PAH emissions is achieved through measures that keep the combustion efficiency as high as possible. Such methods include the increase of excess air, in order to provide oxygen-rich conditions in the freeboard, air staging and an increase in residence time. Other factors influencing PAH formation are oxygen concentration, residence time and air staging. The aim of these measures is to enhance combustion efficiency, thus leading to lower PAHs [39].

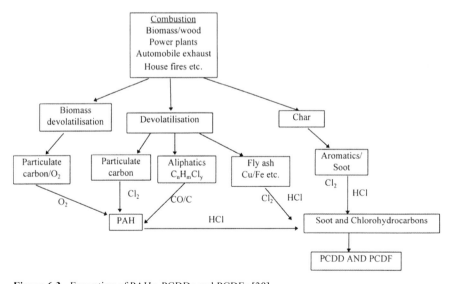

Figure 6.3 Formation of PAHs, PCDDs and PCDFs [38]

6.4.2 PPCD/F

Dioxin is a general term employed for a group of halogenated organic compounds consisting of 75 polychlorinated dibenzo para dioxins (PCDDs) and 135 polycyclic dibenzofurans (PCDFs). They are structurally very similar, comprising of two benzene rings, linked by a single or double oxygen bridge. The only difference is in the number and spatial arrangement of chlorine atoms in the molecule. Seventeen types of Dioxin isomers are highly toxic and suspected to have carcinogenic and mutagenic properties [40]. Due to their lipophylic abilities, dioxins bioaccumulate in living tissue; therefore they can find their way up the food chain and reach dangerous levels even if the initial exposure is small.

PPCD/Fs are always formed in combustion conditions where oxygen, carbon and chlorine are present. The research on the dioxin formation mechanisms is still ongoing. In general, four sources of dioxin formation have been identified [2, 41]:

1. dioxins entering with the feed, *e.g.* in pesticides or wood preservatives;
2. the pyrosynthesis path, *e.g.* the formation of dioxins through homogeneous gas reactions;
3. formation of dioxins through precursors, such as phenols, PAHs, acetylene, lignin and flyash based metallic catalysts;
4. formation through *de novo* reactions, involving carbon bound in fly ash, chlorine and metallic catalysts.

Figure 6.2 illustrates the major dioxin formation mechanisms and their connection to PAH formation. It should be noted that conflicting data has been reported concerning the correlation between PAHs and PPCD/Fs, ranging from simultaneous increase to suppression of dioxin formation due to PAHs and even to noncorrelation [38]. All the above mechanisms are very sensitive to temperatures. For example, dioxins entering in the combustion system with the feed are thermally degraded provided there is sufficient residence time at high temperature zones (at least 2 s at 850°C and 1 s at 1,000°C) [42]. The third and fourth mechanisms are generally considered as the most important, with maximum rates exhibited at 300°C for the De Novo synthesis and 400°C for the precursor path [2]. Dioxins are commonly formed due to reactions on particles entrained in the flue gas in the low temperature zone or deposited on low-temperature surface areas.

Wood combustion is credited with a significant portion of the total dioxin emissions. However, most of the emitted dioxin comes from uncontrolled sources, such as forest fires, or residential applications. In larger facilities, where good combustion conditions prevail and secondary devices for particle removal are present, dioxins can be effectively controlled and kept within the regulatory limits [40].

Dioxin avoidance in fluidized beds is achievable through the control of combustion parameters, for example excess air, secondary/primary air ratio, aiming to increase the overall combustion efficiency [43] and minimize the flue gas residence time in the critical, low temperature zone (250–450°C). The use of secondary emission reduction measures, such as ESPs, bag filters and scrubbers are encouraged. Moreover, the use of additives and inhibitors, such as inorganic S- and N-compounds (*e.g.* SO_2 and NH_3), limestone and urea, are also reported to be effective [44, 45]. Co-combustion of biomass with coal is expected to lead to lower dioxin emissions compared to dedicated biomass combustion, especially in the presence of sulphur in the fossil fuel [3].

6.4.3 CO

Carbon monoxide is the most important intermediate in the conversion path of carbon to CO_2. CO emissions are commonly used as a bench mark for the combustion efficiency of a unit. Carbon monoxide can be produced in significant quantities during biomass combustion, especially in small scale facilities, where the optimization of combustion process is not as thorough, or in units that are were initially designed for coal combustion and have not been retrofitted for biomass firing or co-firing.

Volatile content is the most important biomass property that affects CO formation. As previously noted, biomass fuels have higher volatile contents compared to coals and therefore require a longer residence time in the freeboard to ensure complete combustion of the volatile species. Thus, longer freeboards are required for biomass combustion. However, this feature may not be present in older facilities that were designed for coal combustion. Large particle size and high ash content have also been reported to contribute to CO formation during the char combustion phase [2]. In cases of high ash, the fuel particles follow the shrinking sphere and not the shrinking core model [46], which results in a surrounding ash layer inhibiting the oxygen diffusion to the particle surface.

Temperature is a factor which significantly affects CO emissions. In order to minimize combustion efficiency, the temperature should be kept as high as possible. Therefore, the use of internal heat exchangers in the freeboard of small scale FBCs should be avoided, in order to maintain higher temperatures and ensure the conversion of CO to CO_2. The availability of oxygen is also an important factor towards controlling CO formation. For a given system, there is an ideal excess air ratio to minimize CO emissions; lower values do not ensure adequate oxygen concentration and mixing, while higher values decrease the combustion temperature [1]. Air staging and control of the fluidization velocity to increase the residence time have also been reported as having a positive effect on CO emissions [47]. However, both these measures should be carefully evaluated before implementation, since they could have a negative effect on the boiler performance (*e.g.* reduced load during reduction of fluidization velocity) or even an increase of CO emissions during air staging, due to inappropriate levels of excess air.

6.4.4 NO$_x$

The term nitrogen oxide (NO$_x$) includes both nitrogen oxide (NO) and nitrogen dioxide (NO$_2$). NO$_x$ emissions are considered one of the most important pollutants of combustion systems, due to their role in atmospheric reactions that aggravate particulate matter creation, ground-level ozone and acid rain. In combustion processes, NO$_x$ is formed through three major pathways:

1. thermal NO$_x$ which is formed through high temperature oxidation of the diatomic nitrogen found in combustion air;
2. fuel NO$_x$ which is formed from fuel bound nitrogen;
3. prompt NO$_x$ which is attributed to the reaction of atmospheric nitrogen with fuel bound hydrocarbon radicals.

In FBC systems, the temperature is lower typically lower than 900°C, meaning that the thermal NO$_x$ formation is negligible, while fuel NO$_x$ is responsible for the majority of NO$_x$ emissions [11, 48]. Significant prompt NO$_x$ formation has also been reported by some researchers [49, 50]. The low temperature operation of FBC systems renders thermal NO$_x$ formation obsolete.

The formation of NO$_x$ from fuel bound nitrogen takes place via two different paths, *i.e.* through homogeneous gas-phase reactions of nitrogenous volatile compounds or through heterogeneous oxidation of char-bound nitrogen species. The distribution of nitrogen between volatile compounds and char is roughly proportional to the volatile content of the fuel [2]. NO is the dominant species formed inside the combustion system (> 90%), with NO$_2$ being less than 10%. However, at lower temperatures downstream of the flue gas path and at atmospheric conditions, NO and NO$_2$ are interchangeable [1].

The major nitrogenous volatile species are NH$_3$ and HCN. The formation of HCN is favoured as the rank of the fuel increases. For biomass, an NH$_3$/HCH ratio of 9:1 has been suggested [51] due to the younger age of the fuel. It should be underlined that NH$_3$ has a conversion of 68 and 47% to NO$_x$ whereas HCN has a 90 and 98.4% conversion for bed temperature of 800 and 900°C, respectively [52]. The formation of NH$_3$ is also favoured by the low temperature conditions of FBCs, as has been verified experimentally [53]. Generally, NH$_3$ decomposes to NH$_2$ and NH radicals that in turn can either by oxidized by O$_2$ to form NO or react with available NO and OH radicals to form nitrogen and water vapour. The former reaction typically takes place in the bottom region of FBCs with bottom air injection, while the latter occurs in fuel-rich zones, where the NH$_3$ concentration is increased.

Fuel NO$_x$ formation from biomass combustion is not strongly dependent on temperature, although a decrease of NO$_x$ emissions with decreasing temperatures has been found [1, 3]. The availability of oxygen, fuel reactivity and fuel nitrogen percentage all have an important effect on fuel NO$_x$ formation [2].

Apart from contribution to the NO$_x$ emissions through the oxidization of char-bound nitrogen, char particles are very effective in reducing the overall NO$_x$ emissions by catalyzing the reduction of NO by CO on the fuel particle surface. Sev-

eral other species found in fuel ash, such as CaO, MgO and Fe_2O_3, can also cata-
lyze the reduction of NO to N_2, especially under fuel rich conditions [10]. These
reactions are considered the cause for the declining NO_x formation with increasing
fuel nitrogen content [1, 2]. For coal combustion, about 50% of the gas phase NO_x
reduction is attributed to this heterogeneous catalytic reaction [2]. However, the
effect is less pronounced in biomass fuels, due to the increased volatile and lower
char content. An overview of the fuel nitrogen conversion pathways for biomass
combustion is presented in Figure 6.4 [54].

Although NO_x emissions from biomass facilities appear to be in most cases
within the acceptable limits and less than coal-fired installations, measurements in
laboratory units and large scale plants have confirmed that high NO_x emissions can
be generated during biomass combustion [2, 10]. As a result, both primary and
secondary measures have been investigated. Air and fuel staging are among the
most important primary measures. In staged air combustion, part of the combustion
air is not fed through the bottom of the furnace but rather at a section further downstream. As a result, devolatilization is separated from gas phase combustion. Due
to the scarcity of oxygen in the bottom area, the released NH_3 and HCN in the volatile gases will not react with O_2 towards the formation of NO but rather react with
already released NO towards N_2. A good burnout is ensured by the air influx at the
second stage. A 30–50% NO_x reduction is possible by means of air staging [55].

Fuel staging is another option for NO_x reduction. The majority of the fuel is
combusted with an excess air ratio above 1, which results in the formation of NO_x.
The remainder of the fuel is injected into the flue gas after the primary combustion
zone, without additional air. In the sub-stoichiometric conditions of this zone, NO_x
is reduced to N_2 either through reactions with NH_3 and HCN, as in the case of air
staging, or by reacting with char particles and CH_i radicals [1]. Burnout is
achieved through the addition of air after the reducing zone. Fuel staging requires
lower temperatures than air staging; however the furnace concept and operation is
more complex due to the two independent fuel feeding systems [2]. Reburning is a

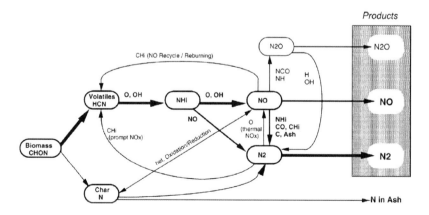

Figure 6.4   Fuel nitrogen conversion pathways during biomass combustion [54]

variation of the fuel staging technique and involves the injection of a secondary fuel in the fuel-rich reducing zone. Both natural gas and wood are considered good reburning fuels [2]. All types of staged combustion require careful optimization and accurate control in order to ensure the required excess air values at each zone.

Secondary measures have also been used in cases where primary measures are inadequate for reducing the emissions at the required levels. The main measures are Selective Catalytic Reduction (SCR) and Selective Non-Catalytic Reduction (SNCR), which inject ammonia or urea as a reducing agent. The difference is that in SCR, due to the utilization of a catalyst, the reduction reaction takes place at a temperature of 250–450°C, while in the SNCR no catalyst is used and the reducing agent is injected at a typical temperature range of 850–950°C. Achieved reductions of up to 95% for SCR and 90% for SNCR have been reported [1], although SNCR typically achieves reductions in the region of 35% at most. Hybrid SNCR/SCR are also utilized. However, issues such as catalyst deactivation, NH_3 slips in the flue gas and unwanted NO formation in non-optimized systems make the adoption of primary measures more preferable.

6.4.5 N_2O

Nitrous oxide is an important greenhouse gas with global warming potential of more than 200 times compared to CO_2. Moreover, it is the major contributor to ozone depletion in the stratosphere. Nitrous oxide emissions are relevant for low temperature combustion applications, such as FBCs, since at temperatures higher than 900°C N_2O decomposes to N_2 [2]. Studies indicate that HCN is a more important precursor for N_2O compared to NH_3 [56]. Therefore, due to the increased NH_3/HCN ration, N_2O emissions are typically very low from biomass combustion applications.

Several N_2O abatement strategies have been suggested, such as increasing the bed temperature, decreasing the excess air, afterburning of a gaseous fuel in the cyclone or freeboard and catalytic reduction of N_2O by metal oxides from the fuel ash [56].

6.4.6 SO_x

Sulphur oxides are formed as a result of the complete oxidation of fuel bound sulphur. SO_2 comprises more than 95% of SO_x emissions, with the remaining part being SO_3 [1]. SO_x are one of the main causes of acid rain; moreover, along with chlorine, they have an important effect on the corrosion chemical pathways inside a boiler. The ability of FBCs to control SO_x emissions through primary measures, *e.g.* the absorption of SO_x in alkali-earth bed material, reduces the installation costs associated with FGD systems and is one of the most attractive characteristics of the technology [57]. Limestone and dolomite are commonly used in FBCs to control sulphur dioxide emissions. The reaction path for limestone is presented in (6.3) and (6.4), which represent the initial endothermic calcination and the subsequent exothermic sulphation reactions. A similar reaction pathway is exhibited by dolomite, which is used instead of limestone in pressurized fluidized bed applications [3].

$$CaCO_3 \rightarrow CaO + CO_2 \tag{6.5}$$

$$CaO + \tfrac{1}{2}O_2 + SO_2 \rightarrow Ca_2SO_4 \tag{6.6}$$

Although the theoretical utilization of limestone be 100% (1 mole of sulphur required 1 mole of calcium), the actual utilization is much lower, ranging from 25 to 45% [2]. The reason for this is the larger volume of Ca_2SO_4 compared to $CaCO_3$. Thus, as the sulphation reaction proceeds, the pore entrances are blocked by the reaction products, leaving the interior without a chance to react [3, 58]. Increased particle size has a negative effect on sulphur retention, due to the larger size of the unreacted core. Optimal reactivity is reached at a temperature of 800–850°C, since higher temperatures lead to rapid sulphation rates and the quick closure of the sorbent pores [3]. Several other parameters also affect the absorption rate and efficiency, either positively (such as high excess air, long residence times, cyclone performance) or negatively (such as high moisture, long term exposure of sorbents to high temperatures) [3, 59].

Generally, the thermal utilization of biomass has a very positive effect on SO_x emissions, due to the low percentage of sulphur in the fuel. Furthermore, biomass ashes typically contain high amounts of alkali earth oxides, such as CaO and MgO, which serve as a natural means of capturing sulphur in combustion conditions, as is illustration in (6.6). Although the quantity of these oxides is small compared to the added sorbents, their efficiency is much higher [3].

6.4.7 Dust

Particulate emissions are an important concern for all biomass combustion technologies. Biomass combustion typically leads to relatively high dust emissions, *i.e.* well above the commonly established limit of 50 mg/m^3 at 11 vol% O_2 [54]. Particulates are commonly characterized either by their aerodynamic particle size as submicrons (less than 1 μm) and supermicrons (greater than 1 μm) or by their origin as products of incomplete combustion (*e.g.* soot, condensable tar and char) and as originating from the inorganic content of the fuel ash.

In the case of CFBs, in contrast to other firing systems, mass size distributions of fly ash particles in FBCs are typically reported to be bimodal, containing both a submicron and a supermicron maximum [2]. Supermicron particles dominate these distributions, with finer particles being only a small part of the particle mass flow. This trend facilitates the efficient dust removal by appropriate filters.

In general, refractory species such as calcium, magnesium, silicon, phosphorus, and aluminium are the main components of the supermicron particles [60, 61] and principally reflect the mineral composition of the fuel. Submicron particles, on the other hand, are mainly composed of alkali salts like potassium chloride and potassium sulphates [62]. Traces of volatile heavy metals can also be found in the aerosol fraction (see Section 6.4.8). Particles concentrations in the flue gas seem to correlate with increased chlorine, and sulphur concentrations in the fuel [62].

As concerns the dust origin, organic particulate emissions are a concern only when the combustion conditions are unsatisfactory; usually, they can be effectively controlled through the change of operating parameters aiming to maximize the combustion efficiency. Due to the favourable combustion conditions of most biomass-fired FBCs, the organic content of dust particles is negligible.

Inorganic particles are formed by two different mechanisms: (1) coarse particles are produced through fusion of non-volatile ash elements in the burning char particles and (2) fine particles are produced from nucleation and condensation of volatile ash elements [63]. The nucleation formation mechanisms result in almost perfect spherical shapes for the submicron particles, while coarse particles in FBCs exhibit irregular shapes, since the temperature is usually quite low to allow the full melting of the ash particles.

Particle removal is achieved through secondary measures aiming to control aerosol precipitation. Cyclone separators are utilized for coarse fly ash (> 5 microns), while electrostatic filters (ESPs) and bag house filters (BHFs) can also be utilized to dump down dust emissions, although their application is limited to medium to large scale units for economic reasons. However, both filter types do not perform satisfactory for submicron particles.

6.4.8 Heavy Metals

The presence of trace elements, such as As, Ba, Cd, Co, Cr, Cu, Hg, Mn, Mo, Ni, Pb, Sb, V and Zn has been confirmed for several biomass fuels, with concentrations varying depending on biomass origin. For agricultural biomass, trace elements are accumulated during the growth stage from external sources, such as polluted soil and fertilizers. As a result, their type and concentration varies with pollutant type, distance from the source and plant age. For industrial biomass residues, the presence of trace elements is mostly dependent on past processing and utilization. As a result, the concentration of trace elements in different biomass species, or even in different samples of the same variety, can vary by a factor of up to 100. For example, values of 60–640 mg/kg have been reported for Zn, and a range of 0.1–6.6 mg/kg for Cd [64].

Heavy metals partitioning to the different ash categories depends upon many factors, such as temperature and air distribution in the bed, as well as on chemical and physical properties of the respective fuel. Furthermore, the partitioning is also influenced by the particle size distribution of the fly ash and the gaseous concentrations in the surrounding gas [65]. A relatively small fraction of the overall trace element concentration remains in the bed, while a larger fraction is entrained along with the fly-ash.

Non-volatile elements, such as Fe, Cr, Cu and Al, have the tendency to form stable oxides and accumulate in coarser ash particles, which are retrieved in the cyclone. Since these elements have important nutritional value in agriculture, biomass cyclone ash has the potential to be utilized as a fertilizer in fields [2].

The majority of the volatile elements end up in fine ash particles, especially in the submicron range, since their large surface area acts as a sink for metal vapours [66]. As a result, significant amounts of volatile heavy metals escape in fine dust form along with the stack gases. Chlorine, as in the case of alkalis, also facilitates the transfer of the volatile trace metals (Cd, Pb, Ar, and Zn) in the fly ash in the form of chlorides and oxides [2]. Cd and Pb are preferentially converted into $CdCl_2/PbCl_2$ during combustion. Zn can also be volatilized as a chloride; however, due to the formation of a stable oxide form, a significant amount remains in the fuel ash and is recovered in the cyclone. Ar and Sb exhibit similar behaviour, though their concentration is typically low. Hg is also highly volatile; however its concentration in untreated biomass fuels is much lower than most coals, ranging from 0.01 to 0.1 mg/kg d.b. [65, 67, 68].

In industrial applications, the trace elements emissions are mostly controlled through secondary measures for minimizing dust emissions. However, primary measures have also been suggested, aiming to use adsorbent materials, such as alumina, kaolinite, bauxite and emathlite, for suppressing the volatilization of heavy metals and retaining them in the fuel ash [60].

6.5 Conclusions

Given the growing demand for reduction of CO_2 emissions, biomass is considered to be one of the most important sources for renewable energy production. Several different technologies have been developed or adapted for biomass fuels; however, the fluidized bed combustion technology remains one of the most popular due to its ability to alleviate many of the operational or environmental issues that arise from the biomass fuel properties.

Different biomass fuels have varying chemical or physical properties, which renders their utilization in standard combustion facilities, custom-built for specific fuel qualities, difficult. However, FBCs are well equipped to deal with such changes of fuel quality. Moreover, they are less sensitive to feeding blockages due to the high moisture content of biomass. The high volatile content of the fuel is also not an issue, since staged combustion is easily implemented.

Pollutants from incomplete combustion, such as CO and dioxins, are usually not an issue due to the high combustion efficiency of FBCs. SO_x emissions are typically very low due to the low sulphur content of biomass, while NO_x emissions can be handled through primary or secondary measures. N_2O emissions, which are a typical problem of fossil fuel fired FBC installations, are not an issue for biomass combustion. An additional advantage of FBCs is that many pollutants can be controlled through the correct choice of bed materials or additives. Research to extend these primary measures for some problematic environmental aspects of biomass combustion, such as dust and trace element emissions which are only controlled now through secondary measures, is ongoing.

The most significant technical issue concerning biomass utilization in FBCs is the ash and chlorine content. In contrast to coal, the issue is not ash quantity but quality and most importantly its alkali content. Alkali oxides or salts have the tendency to form eutectic mixtures, which are the cause of fouling, corrosion and agglomeration problems, as well as a contributor to the increased heavy metal content in the dust emissions. Since removal of these compounds from the biomass fuel is not yet an efficient option, other measures are currently utilized, such as temperature control and the use of additives. Research initiatives for novel bed material and additives are still ongoing.

A final issue of biomass utilization is the high energy and economic cost associated with drying the fuel and reducing its size. However, FBCs are advantageous, since they can handle high moisture fuels and require much larger particle sizes compared to pulverized fuel combustors.

References

1. Van Loo S, Koppejan J (2008) Handbook of biomass combustion and co-firing. Earthscan, London
2. Khan AA, de Jong W, Jansens PJ, Spliethoff H (2009) Biomass combustion in fluidized bed boilers: potential problems and remedies. Fuel Process Technol 90(1):21–50
3. Prabir B (2006) Combustion and gasification in fluidized beds. CRC Press, Taylor & Francis Group
4. Werther J (1992) Fluidized-bed reactors. Ulmann's encyclopedia of industrial chemistry. Weinheim, VCH, pp 239–274
5. Anthony EJ (1995) Fluidized-bed combustion of alternative solid fuels — status, successes and problems of the technology. Prog Energ Combust 21(3):239–268
6. Leckner B (1998) Fluidized bed combustion: mixing and pollutant limitation. Prog Energ Combust 24(1):31–61
7. Leckner B (1992) Optimization of emissions from fluidized-bed boilers. Int J Energ Res 16(5):351–363
8. Skodras G, Grammelis P, Basinas P, Kakaras E, Sakellaropoulos G (2006) Pyrolysis and combustion characteristics of biomass and waste-derived feedstock. Ind Eng Chem Res 45(11):3791–3799
9. Grammelis P, Malliopoulou A, Basinas P, Danalatos N (2008) Cultivation and characterization of Cynara cardunculus for solid biofuels production in the Mediterranean region. Int J Mol Sci 9(7):1241–1258
10. Werther J, Saenger M, Hartge EU, Ogada T, Siagi Z (2000) Combustion of agricultural residues. Prog Energ Combust 26(1):1–27
11. Demirbas A (2004) Combustion characteristics of different biomass fuels. Prog Energ Combust 30(2):219–230

12. Fahlstedt I, Lindman EK, Lindberg T, Anderson T (1997) Co-firing of biomass and coal in a pressurized fluidized bed combined cycle: results of pilot scale studies. In: Preto FDS (ed) 14th International Conference on Fluidized Bed Combustion. ASME, New York

13. Demirbas A (2001) Relationships between lignin contents and heating values of biomass. Energ Convers Manage 42(2):183–188

14. Nielsen HP, Frandsen FJ, Dam-Johansen K, Baxter LL (2000) The implications of chlorine-associated corrosion on the operation of biomass-fired boiler. Prog Energ Combust 26(3):283–298

15. Baxter L (2005) Biomass-coal co-combustion: opportunity for affordable renewable energy. Fuel 84(10):1295–1302

16. Tillman DA (2000) Biomass cofiring: the technology, the experience, the combustion consequences. Biomass Bioenerg 19(6):365–384

17. Baxter LL (1993) Ash deposition during biomass and coal combustion – a mechanistic approach. Biomass Bioenerg 4(2):85–102

18. Baxter LL (1998) Influence of ash deposit chemistry and structure on physical and transport properties. Fuel Process Technol 56(1/2):81–88

19. Wei X, Schnell U, Hein KRG (2005) Behaviour of gaseous chlorine and alkali metals during biomass thermal utilisation. Fuel 84(7/8):841–848

20. Coda B, Aho M, Berger R, Hein KRG (2001) Behavior of chlorine and enrichment of risky elements in bubbling fluidized bed combustion of biomass and waste assisted by additives. Energ Fuel 15(3):680–690

21. Dayton DC, Belle-Oudry D, Nordin A (1999) Effect of coal minerals on chlorine and alkali metals released during biomass/coal cofiring. Energ Fuel 13(6):1203–1211

22. Aho M (2001) Reduction of chlorine deposition in FB boilers with aluminium-containing additives. Fuel 80(13):1943–1951

23. Lin W, Dam-Johansen K, Frandsen F (2003) Agglomeration in bio-fuel fired fluidized bed combustors. Chem Eng J 96(1/3):171–185

24. Lin CL, Kuo JH, Wey MY, Chang SH, Kai-Sung Wang (2009) Inhibition and promotion: the effect of earth alkali metals and operating temperature on particle agglomeration/defluidization during incineration in fluidized bed. Powder Technol 189(1):57–63

25. Bapat DW, Kulkarni SV, Bhandarkar VP (1997) Design and operating experience on fluidized bed boiler burning biomass fuels with high alkali ash. In: Preto FDS (ed) 14th International Conference on Fluidized Bed Combustion. ASME, New York

26. Grubor BD, Oka SN, Ilic MS, Dakic DV, Arsic BT (1995) Biomass FBC combustion-bed agglomeration problems. In: Heinschel KJ (ed) Proceedings of the 13th International Conference on Fluidized Bed Combustion. ASME, New York

27. Lin L, Gitte K, Kim DJ, Esther M, Bank L (1997) Agglomeration phenomena in fluidized bed combustion of straw. In: Preto FDS (ed) 14th International Conference on Fluidized Bed Combustion. ASME, New York

28. Scala F, Chirone R (2006) Characterization and early detection of bed agglomeration during the fluidized bed combustion of olive husk. Energ Fuel 20(1):120–132

29. Scala F, Chirone R (2008) An SEM/EDX study of bed agglomerates formed during fluidized bed combustion of three biomass fuels. Biomass Bioenerg 32(3):252–266

30. Silvennoines J (2003) A new method of inhibiting bed agglomeration problems in fluidized bed boilers. In: Pisupati S (ed) Proceedings of the 17th International Conference on Fluidized Bed Combustion. ASME, New York

31. Glazer MP, Schurmann H, Monkhouse P, Jong de W, Spliethoff H (2005) Co-combustion of coal with high alkali straw: measuring of gaseous alkali metals and sulfur emissions monitoring. In: Cen K (ed) Circulating fluidized bed technology – VIII. International Academic Publishers, Beijing

32. Yrjas P, Skrifvas BJ, Hupa M, Reppo J, Nylund MP, Vaimikku P (2005) Chlorine in deposits during co-firing of biomass peat and coal in full scale CFBC boiler. In: Jia L (ed) Proceedings of the 18th International Conference on Fluidized Bed Combustion. ASME, New York

33. Bolhŕr-Nordenkampf M, Gartnar F, Tschanun I, Kaiser S (2008) Operating experiences from two new biomass fired FBC-plants with a high fuel flexibility and high steam parameters. In: Werther J, Nowak W, Wirth KE, Hartge EU (eds) Proceedings of the 9th International Conference on Circulating Fluidized Beds. TuTech Innovation GmbH, Hamburg

34. Seemann B, Steer T, Hein D (2008) Consistent further development of the BFB-Technology – application of the BioCOM® for problematic residues of the wood and fibreboard industry. In: Werther J, Nowak W, Wirth KE, Hartge EU (eds) Proceedings of the 9th International Conference on Circulating Fluidized Beds. TuTech Innovation GmbH, Hamburg

35. Simoneit BRT (2002) Biomass burning – a review of organic tracers for smoke from incomplete combustion. Appl Geochem 17(3):129–162

36. Mastral AM, Callen MS (2000) A review an polycyclic aromatic hydrocarbon (PAH) emissions from energy generation. Environ Sci Technol 34(15):3051–3057

37. Mastral AM, Callen MS, Garcia T (2000) Fluidized bed combustion (FBC) of fossil and nonfossil fuels. A comparative study. Energ Fuel 14(2):275–281

38. Chagger HK, Kendall A, McDonald A, Pourkashanian M, Williams A (1998) Formation of dioxins and other semi-volatile organic compounds in biomass combustion. Appl Energ 60(2):101–114

39. Liu K, Han W, Pan WP, Riley JT (2001) Polycyclic aromatic hydrocarbon (PAH) emissions from a coal-fired pilot FBC system. J Hazard Mater 84(2/3):175–188

40. Lavric ED, Konnov AA, DeRuyck J (2004) Dioxin levels in wood combustion – a review. Biomass Bioenerg 26(2):115–145

41. Gulyurtlu I, Crujeira AT, Abelha P, Cabrita I (2007) Measurements of dioxin emissions during co-firing in a fluidised bed. Fuel 86(14):2090–2100

42. McKay G (2002) Dioxin characterisation, formation and minimisation during municipal solid waste (MSW) incineration: review. Chem Eng J 86(3):343–68

43. Ishikawa R, Buekens A, Huang H, Watanabe K (1997) Influence of combustion conditions on dioxin in an industrial-scale fluidized-bed incinerator: experimental study and statistical modelling. Chemosphere 35(3):465–477

44. Zhong ZP, Jin BS, Huang YJ, Zhou HC, Lan E (2006) Experimental research onemission and removal of dioxins in flue gas from a co-combustion of MSWand coal incinerator. Waste Manage 26(6):580–586

45. Ruokojärvi PH, Asikainen AH, Tuppurainen KA, Ruuskanen J (2004) Chemical inhibition of PCDD/F formation in incineration processes. Sci Total Environ 325(1/3):83–94

46. Basu P (1999) Combustion of coal in circulating fluidized-bed boilers: a review. Chem Eng Sci 54(22):5547–5557

47. Khan AA, Aho M, deJong W, Vainikka P, Jansens PJ (2008) Scale-up study on combustibility and emission formation with two biomass fuels (B quality wood and pepper plant residue) under BFB conditions. Biomass Bioenerg 32(12):1311–1321

48. Leckner B, Karlsson M (1993) Gaseous emissions from circulating fluidized-bed combustion of wood. Biomass Bioenerg 4(5):379–389

49. Miller JA, Bowman CT (1998) Mechanism and modeling of nitrogen chemistry in combustion. Prog Energ Combust 15(4):287–338

50. Nordin A (1994) Chemical elemental characteristics of biomass fuels. Biomass Bioenerg 6(5):339–347

51. Liu H, Gibbs BM (2002) Modeling of NO and N_2O emissions from biomass-fired circulating fluidized bed combustors. Fuel 81(3):271–280

52. Desroches-Ducarne E, Dolinger JC, Marty E, Martin G, Delfose L (1998) Modelling of gaseous pollutants emissions in circulating fluidized bed combustion of municipal refuse. Fuel 77(13):1399–1410

53. Johnsson JE, Amand LE, DamJohansen K, Leckner B (1996) Modeling N_2O reduction and decomposition in a circulating fluidized bed boiler. Energ Fuel 10(4):970–979

54. Nussbaumer T (2003) Combustion and co-combustion of biomass: fundamentals, technologies, and primary measures for emission reduction. Energ Fuel 17(6):1510–1521

55. Obernberger I (1998) Decentralized biomass combustion: state of the art and future development. Biomass Bioenerg 14(1):33–56
56. Mann MD, Collings ME, Botros PE (1992) Nitrous-oxide emissions in fluidized-bed combustion – fundamental chemistry and combustion testing. Prog Energ Combust 18(5):447–461
57. Oka SN (2004) Fluidized bed combustion. Marcel Decker Inc., New York
58. Mulligan T, Pomeroy M, Bannard JE (1989) The mechanism of the sulfation of limestone by sulfur-dioxide in the presence of oxygen. J I Energ 62(450):40–47
59. Laursen K, Grace JR (2002) Some implications of co-combustion of biomass and coal in a fluidized bed boiler. Fuel Process Technol 76(2):77–89
60. Lind T, Valmari T, Kauppinen EI, Sfiris G, Nilsson K, Maenhaut W (1999) Volatilization of the heavy metals during circulating fluidized bed combustion of forest residue. Environ Sci Technol 33(3):496–502
61. Kaufmann H, Nussbaumer T (1999) Formation and behaviour of chlorine compounds during biomass combustion. Gefahrst Reinhalt L 59(7/8):267–272
62. Johansson LS, Tullin C, Leckner B, Sjovall P (2003) Particle emissions from biomass combustion in small combustors. Biomass Bioenerg 25(4):435–446
63. Wiinikka H, Gebart R, Boman C, Boström D, Öhman M (2007) Influence of fuel ash composition on high temperature aerosol formation in fixed bed combustion of woody biomass pellets. Fuel 86(1/2):181–193
64. Demirbas A (2005) Potential applications of renewable energy sources, biomass combustion problems in boiler power systems and combustion related environmental issues. Prog Energ Combust 31(2):171–192
65. Obernberger I (1997) Concentrations of inorganic elements in biomass fuels and recovery in the different ash fractions. Biomass Bioenerg 12(3):211–224
66. Obernberger I, Brunner T, Barnthaler I (2006) Chemical properties of solid biofuels – significance and impact. Biomass Bioenerg 30(11):973–982
67. Demirbas A (2004) Trace element concentrations in ashes from various types of Lichen biomass species. Energ Source 26(5):499–506
68. Demirbas A (2005) Heavy metal contents of fly ashes from selected biomass samples. Energ Source 27(13):1269–1276

Chapter 7
Gasification Technology and Its Contribution to Deal with Global Warming

Filomena Pinto, Rui André, Paula Costa, Carlos Carolino, Helena Lopes and I. Gulyurtlu

Abstract It is predictable that energy demand will greatly increase in years to come, due to the continuous growth of world population, together with the quest to improve living standards. CO_2 emissions are hence expected to increase significantly. Gasification is a mature technology for energy production that permits an easier separation of CO_2 for its storage. As modern societies are producing ever-increasing amounts of wastes with negative impact on the environment, new technologies have been developed to co-gasify these wastes either with coal or alone, thus resolving a serious problem of waste disposal. Wastes gasification reduces the dependence on fossil fuels and co-gasification with coal could provide the benefit of security in fuel supply, as the availability of wastes and biomass fuels could vary from region to region and show seasonal changes. Gasification experimental conditions and technologies and syngas cleaning methods are key issues for the production of a clean gas that could find a wide range of applications. This chapter will concentrate on syngas end-uses, focusing on new ones, like gas turbines or engines in IGCC, synthesis of methanol, ethanol and dimethyl ether, Fischer–Tropsch synthesis, and hydrogen production. The role of gasification in CO_2 sequestration will also be discussed.

7.1 Introduction

It is predictable that energy demand will have greatly increase in years to come, due to the continuous growth of world population and also because of increasing high quality standards. Energy demand growth, especially from large emerging economies like India and China, may cause substantial alterations in actual world

F. Pinto (✉)
LNEG, Estrada do Paço do Lumiar, 22, 1649-038 Lisboa, Portugal
Tel: 351 21 092 4787
e-mail: filomena.pinto@lneg.pt

organization by raising the price of fossil fuels and by increasing pollutants emissions. To face these challenges and to decrease their impact, it is important to find alternative energy resources. The European Union aims to increase the contribution of renewable energy sources to reach a level of 20% global energy share by 2020 [1]. Most European and USA countries have encouraged the use of renewable energies and the use of biomass and wastes for energy production. On the other hand, fossil fuels will continue to have the greatest share in energy production at least till 2030 with consequent large releases of carbon dioxide (CO_2) emissions, according to IEA predictions [2].

Combustion is the most used technology for energy production, either from fossil fuels or from wastes; however, despite the great efforts to develop new technologies to decrease pollutants emissions from combustion, the release of CO_2 is still huge and may put life on Earth in danger in the near future. Therefore, the development of new technologies suitable for wastes processing and that may also decrease CO_2 emissions to prevent global warming is urgent. Gasification may have an important role in achieving these objectives in the near future, as it allows an easier separation of CO_2 for its sequestration.

Coal gasification is a well known technology but the already available knowledge must be used to research and develop new processes for co-gasification of coal mixed with different types of biomass and wastes with negative impact on the environment. This allows avoiding the serious problems related to waste disposal and the dependence on fossil fuels will be decreased by using alternative fuels.

On the other hand, biomass gasification may also have an important task in attaining European Union objectives for decreasing greenhouse gas emissions, as during biomass growth in a sustainable basis the CO_2 released during biomass utilization may be mostly absorbed. Despite recent research studies on this subject, cleaning technologies for gasification gas (syngas) are under development and further demonstration of these technologies is still needed.

Nowadays, gasification is deployed all over the world for processing mainly coal and petroleum residuals, like petroleum coke (petcoke). Most installations are in Western Europe, the Pacific Rim, Africa, and North America. Besides the advances of biomass gasification, the success of the gasification process depends on the development of cost-effective and technologically viable gas cleaning technologies, mainly when fuels with considerable amounts of S, Cl, and N are used, which, during gasification, may form several undesirable compounds, some of which may be released into the gas phase. The main gasification drawbacks are related to pollutant emissions, production of inert solid residues, higher product flexibility, and higher efficiency to power, and all of these problems need to be solved. Furthermore, more efficient biomass gasification technologies like Integrated Gasification Combined Cycle (IGCC) need to be fully demonstrated.

Gasification is also a promising process to achieve pre-combustion CO_2 capture for storage. Coal and/or wastes are first gasified to produce syngas. After syngas cleaning, CO is converted into more H_2 and CO_2 in the presence of steam by water-shift reaction. The gas produced contains almost exclusively H_2 and CO_2, whose concentration is much higher than in conventional processes. This feature, together with the fact that the gas is at high pressure with almost no impurities, facilitates the CO_2 separation process. Pressure swing adsorption, membrane, or cryogenic separation are the most promising technologies to produce a H_2-rich fuel that can be used in a gas turbine combined cycle or in fuel cells to produce electricity and a CO_2 rich stream ready for storage. Therefore, it is expected that, in the near future, gasification technology may contribute to the production of the energy necessary to ensure economic development and to effective CO_2 sequestration, in order to preserve life on Earth as we know it.

7.2 Gasification Fundamentals

Gasification of solid carbonaceous fuels is a very old technology, which has been used on the industrial scale for coal gasification since the seventeenth century. Biomass gasification and pyrolysis was used to produce solid fuel, while the liquid volatiles (tar) had diverse uses like embalming, meat packing, and wood waterproofing in boats in Ancient Egypt and in Greek, Roman, Chinese and other early civilizations both in the Euro-Asian region and in the Americas. During the first half of the twentieth century and up to World War II, several gaseous and liquid fuels were obtained from coal through gasification, especially where petroleum was not available.

Gasification is a thermochemical process that converts carbonaceous materials, usually coal and/or wastes, either used alone or mixed with one another, into a syngas, whose major components are carbon monoxide and dioxide (CO and CO_2), hydrogen (H_2), methane (CH_4), and other gaseous hydrocarbons, usually referred as C_nH_m. Gasification commonly uses temperature values higher than 750°C up to 1,300°C, depending on the gasification process used. During carbonaceous materials heating, different processes may occur, namely: (1) *drying –* release of water and of some of the more volatile components at temperature values around 110°C; (2) *devolatilization or pyrolysis –* release of volatile compounds, mainly H_2, CO, CO_2, H_2O, CH_4, other gaseous hydrocarbons, NH_3, H_2S, and phenols, and formation of char, which is mostly carbon and the mineral matter of carbonaceous materials at temperature values around 350 C; (3) *gasification –* at temperature values higher than 350°C chemical reactions occur between the carbonaceous materials or the char and the chemical species present in the surrounding medium, which include those released during devolatilization and those supplied to the gasification medium, usually air or oxygen, carbon dioxide and/or steam, or a mixture of some of these compounds. Depending on the type of gasification reactor used these processes may occur at different stages, like in fixed bed reactors or almost simultaneously as happens in fluidized beds.

A wide range of chemical reactions may occur during gasification; the more important ones are summarized in Table 7.1.

Air or oxygen added to the gasification medium promotes the oxidation reactions (7.1)–(7.3), supplies the heat necessary for the endothermic reactions, and releases H_2O, CO and CO_2 for gasification reactions. CO_2 may react with solid carbon through the Boudouard reaction (7.4) to release CO. CO_2 may also participate in dry reforming reactions (7.13)–(7.15) to decompose hydrocarbons into H_2 and CO. H_2O may react with solid carbon by water gas endothermic reactions (7.5) and (7.6) to form H_2, CO, and CO_2 and may also participate in water gas shift (WGS) reaction (7.8) to convert CO into H_2 and CO_2; this reaction is therefore very important to change syngas CO/H_2 ratio. H_2O also takes part in steam reforming reactions (7.9)–(7.12), which convert hydrocarbons into H_2O, CO, and CO_2. Hydrocarbons may also be converted by cracking reactions.

On the other hand, CH_4 may be formed by methanation or hydro-gasification, reaction (7.7) that occurs between carbon and hydrogen. Although it is usually very slow, it may be favored by higher pressure. Hydrogen reforming reaction (7.16) may also form CH_4, the reaction between H_2 and CO usually occurring at low temperature, but is favored by higher pressure or in presence of suitable catalysts. As there are many different reactions occurring during gasification, the products of some of them being the reactants of others, it is difficult to understand fully the complex reactions that may occur and thus predict syngas composition.

Syngas heating value depends on the gasification medium used. Though the use of air has the advantage of supplying the heat necessary for gasification endothermic reactions, the syngas produced has a low heating value, usually 3.5–7 MJ/m^3, due to nitrogen diluting effect. The use of oxygen, instead of air, allows overtaking the diluting effect of nitrogen, thus syngas has a higher heating value, usually 9–15 MJ/m^3, but due to the oxygen production cost, operating costs are higher. Other gasification systems use only steam as gasification medium, which also overtakes the diluting effect of nitrogen and, though steam production is cheaper

Table 7.1 Most important gasification reactions

Designation	Mechanism	ΔH (kJ/mol)	
	$C_{(s)} + O_2 \leftrightarrows CO_2$	−392.5	(7.1)
Oxidation	$C_{(s)} + \frac{1}{2}O_2 \leftrightarrows CO$	−110.5	(7.2)
	$H_2 + \frac{1}{2}O_2 \leftrightarrows H_2O$	−242.0	(7.3)
Boudouard	$C_{(s)} + CO_2 \leftrightarrows 2CO$	172.0	(7.4)
Water Gas: primary	$C_{(s)} + H_2O \leftrightarrows CO + H_2$	131.4	(7.5)
secondary	$C_{(s)} + 2H_2O \leftrightarrows CO_2 + 2H_2$	90.4	(7.6)
Methanation	$C_{(s)} + 2H_2 \leftrightarrows CH_4$	−74.6	(7.7)
Water-gas shift	$CO + H_2O \leftrightarrows CO_2 + H_2$	−41.0	(7.8)
	$CH_4 + H_2O \leftrightarrows CO + 3H_2$	205.9	(7.9)
Steam reforming	$CH_4 + 2H_2O \leftrightarrows CO_2 + 4H_2$	164.7	(7.10)
	$C_nH_m + nH_2O \leftrightarrows nCO + (n+m/2)H_2$	210.1	(7.11)
	$C_nH_m + n/2H_2O \leftrightarrows n/2CO + (m-n)\,H_2 + n/2CH_4$	4.2	(7.12)
	$CH_4 + CO_2 \leftrightarrows 2CO + 2H_2$	247.0	(7.13)
CO_2 reforming	$C_nH_m + nCO_2 \leftrightarrows 2nCO + m/2\ H_2$	292.4	(7.14)
	$C_nH_m + n/4CO_2 \leftrightarrows n/2CO + (m-3n/2)H_2 + (3n/4)CH_4$	45.3	(7.15)
H_2 reforming	$CO + 3H_2 \leftrightarrows CH_4 + H_2O$	−205.9	(7.16)

than oxygen, the heat necessary for gasification process needs to be supplied by external media, either by an inert material like sand (Battelle process) or by a hot fluid circulating through a heat exchanger placed inside the gasifier. Both processes also increase gasifier operating costs.

The production of syngas through gasification is always associated with the release of tar. Tar is a complex mixture of high molecular weight hydrocarbons that may contain different compounds from single ring to five-ring aromatic compounds together with other hydrocarbons containing oxygen and complex polycyclic aromatic hydrocarbon (PAH). The presence of tar may cause several problems. namely sticky deposits that may cause blocking of pipes, gas coolers, filter elements, engine suction elements and tar may also deposit on catalyst surface, deactivating the catalyst that may be used in a downstream process, like: steam reforming, water gas shift reaction, and chemical synthesis. Tar deposition in gas turbines, engines, or boilers causes severe mechanical damage, and therefore most syngas end-uses require very low tar contents, as shown in Table 7.2. Tar abatement is a key issue in gasification and syngas cleaning processes. Several authors have been developing different types of catalysts with the aim of getting a catalyst with high performance and low cost. The catalysts used so far may be divided into four groups: (1) natural minerals (limestone, dolomite, olivine), (2) alkali metals (KOH, $KHCO_3$, and K_2CO_3), (3) Ni-based and (4) noble metal catalysts (Rh, Ru, Pt, and Pd). Some of these catalysts have been tested inside the gasifier by Pinto et al. [3] and/or in a secondary reactor for syngas cleaning, leading the second option for higher tar abatement by Pinto et al. [4]. Further research and development is still required till more effective catalysts for tar abatement are found, the main challenges being the development of catalysts with longer lifetimes, higher mechanical strength, low cost, and high tar decomposition capacity.

Besides the mentioned syngas components, other undesirable compounds may also be formed, especially when low grade coals or wastes are gasified. When carbonaceous materials with high contents of N, S, and halogens are gasified, the formation of NH_3, H_2S, and HCl is expected, due to the reducing gasification conditions. These compounds are pollutant precursors as they originate NO_x and SO_x when syngas is burned for energy production. On the other hand, H_2S may also poison catalysts used in syngas cleaning processes.

The formation of NH_3, H_2S, and HCl may be controlled through the adjustment of gasification operating parameters, such as size, shape, structure and mineral composition of carbonaceous materials, gasification medium, temperature and heating rate, and the use of specific catalysts or sorbents by Pinto et al. [5]. These parameters also affect gasification performance and syngas properties and composition.

Usually syngas produced by gasification does not have the suitable characteristics required by its utilizations so gas conditioning is necessary to decrease to low levels the contents of undesirable compounds such as tar, H_2S, HCl, and NH_3. The presence of these compounds increases operational costs, since they poison and deactivate catalysts, but also promote and increase corrosion in the equipment. On the other hand, as they are pollutant precursors, the presence of such contaminants is environmentally adverse and must comply with emission limits legislation.

Therefore, the reduction of these compounds' contents is fundamental. When the syngas is going to be used at atmospheric and temperature conditions a possible option for syngas conditioning is wet scrubbing. However, if the syngas is going to be used in any thermal process, such as shift reactor or chemical synthesis, hot gas cleaning processes are the best option, as they allow higher energy efficiency and avoid the production of wastewater contaminated with tar, inorganic acids, NH_3, and metals.

To sum up, it may be said that the most efficient way to remove tar, NH_3, H_2S, and HCl is by several steps, beginning with a cheaper material that removes the larger fraction of H_2S and HCl, being followed by a more specific catalyst that may remove the remaining tar. Pinto et al. [4] tested a configuration with two catalytic fixed bed reactors, the first one with dolomite to retain H_2S and HCl and to promote some tar decomposition and the second one with a more expensive Ni-based catalyst to eliminate completely tar and gaseous hydrocarbons, apart from methane. This configuration proved to be suitable to treat syngas with high contents of tar, sulfur and halogen compounds, which by being retained in dolomite reactors will ensure a longer life for Ni-based materials used in the second reactor. This configuration could be simplified by suppressing the dolomite reactor whenever the syngas has low H_2S and HCl contents or by omitting the Ni-based catalyst reactor when either the syngas has low tar contents or its application is not very exigent towards the existence of tar.

7.3 Syngas Utilizations

7.3.1 Introduction

Syngas may have a large range of end-uses. However, most of them are very demanding towards syngas quality and impurities contents, as summarized in Table 7.2, which obliges a more or less complex syngas cleaning process, depending on syngas composition. In Figure 7.1 one possible configuration is presented for hot syngas conditioning, which may be simplified depending on the type and composition of carbonaceous materials gasified, gasification conditions, and addition of catalysts or sorbents. In Figure 7.1 there are also presented syngas applications in different conversion processes and the main product obtained in each one.

Nowadays, the production of heat and power has two main challenges: the increase of processes efficiency and the minimization of green house gas emissions. Due to the large reserves of coal and to the possibility of using it mixed with wastes, the use of syngas for heat and power production has grown in importance and interest. Syngas may be used in boilers or combustors for heat production, which are less demanding towards syngas quality and characteristics than some other syngas utilizations like motors or turbines, since the presence of small amounts of tar and other impurities are allowed in syngas, as shown in Table 7.2.

Usually the flue gas produced by an IGCC system is fed to a gas turbine, fuel cell, and steam turbine for power generation, and a steam generator for heat recovery. In a gas turbine, due to the formation of alkali sulfates, the H_2S concentration must be limited to less than 20 ppm to protect it from high temperature corrosion.

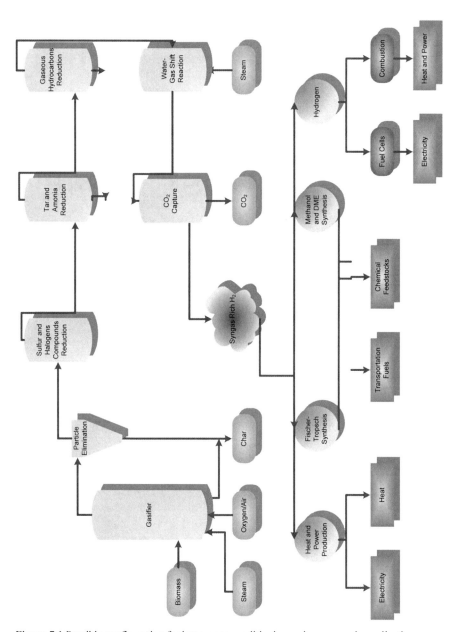

Figure 7.1 Possible configuration for hot syngas conditioning and syngas main applications

Table 7.2 Syngas characteristics and quality demanded by different utilizations

Impurity	Boiler	Gas engines	Gas turbines	Fuel cells	Chemical synthesis
Particulate (mg/Nm^3)	1,000	<50	<15	<0.1	Almost completely removed
Particle size (µm)	10	<10	<5	<10	
Tar (Dew point)	<400°C	<10°C		–	Not condensing below dew point
Alkali metals		0.24 mg/Nm^3	0.24 mg/Nm^3	<10 ppm	10 ppbv
NH_3	–	<50 mg/Nm^3		<5000 ppm	1 ppmv
Total sulfur	72 mg/Nm^3	<80 mg/Nm^3		<1 ppm	<1 ppmv
Total chlorine	35 mg/Nm^3	<100 mg/Nm^3		<1 ppm	10 ppbv

The Ni catalyst and anode of the fuel cells are poisoned by H_2S, resulting in loss of cell voltage; therefore its concentration must be reduced to under 1 ppm as reported by Ohtsuka et al. [6].

Due to the low heating value of syngas produced when air is used as gasification medium (3.5 MJ/m^3 and 7 MJ/m^3), syngas transport and storage is not economical viable. Therefore, syngas should be burned near the gasifier to decrease heat losses and to guarantee a high global efficiency. When gasification takes place in the absence of air, syngas has a medium heating value, usually 9–15 MJ/m^3, which allows its transportation, storage, and utilization in different applications.

7.3.2 Heat and Power Production Through Engines and Turbines

When the syngas has low contents of particulates and tar it can be used in gas turbines or in motors as far as it fits the requirements of Table 7.2. In gas turbines, syngas chemical energy is converted into mechanical energy, which is used to produce electricity. In engines syngas is burned and converted into CO_2 and H_2O. Syngas has been used in engines, though, as reported by Sridhar et al. [7], the combustion chamber should be modified in relation to that used for diesel, to decrease energy losses and increase efficiency.

The remaining heat of the exhaust gas that leaves the turbine, the engine, or the boiler may be used to produce steam which, when used in a steam turbine to produce additional electricity, allows increasing of the energy conversion efficiency of the overall process, as shown in Figure 7.2. This concept is used in IGCC, which are great energy management installations and may be considered to be a relatively environmentally friendly method of using coal, especially when they also integrate CCS (carbon capture and sequestration) units to decrease CO_2 emissions. IGCC installations may present different configurations, incorporating gasification units with gas cleaning processes and power production units with engines and/or gas and steam turbines. There are commercial and demonstration

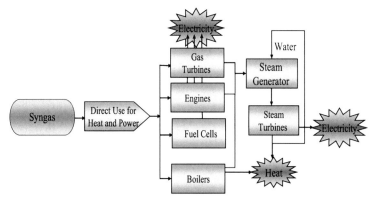

Figure 7.2 Direct use of syngas for heat and power production

IGCC installations for power generation from natural gas or from syngas produced by coal gasification in the United States, Europe and Japan.

In Figures 7.3 and 7.4 are presented general configurations for IGCC installations using engines or turbines, respectively. In Figure 7.4 is also considered an air separation unit to produce oxygen for the gasifier, using the nitrogen stream to introduce inside the combustor, because Frey *et al.* [8] analyzed different integration possibilities by simulation using ASPEN Plus with and without air extraction and nitrogen injection and combinations of both at different pressures and concluded that nitrogen injection at elevated pressure led to high efficiency and to low emissions.

Mondol *et al.* [9] used the software ECLIPSE to evaluate the techno-economic performance of four new concepts for IGCC with CO_2 capture facilities, which

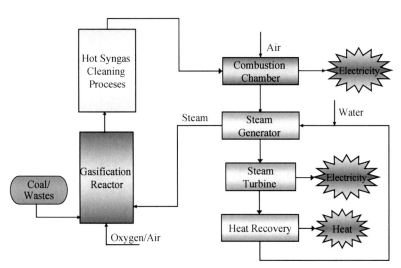

Figure 7.3 Schematic diagram for general configuration of IGCC installation, combining the gasifier and syngas cleaning systems with an internal combustion engine

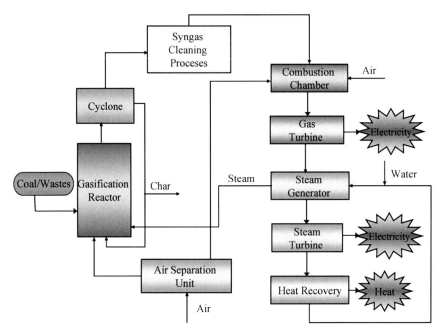

Figure 7.4 Schematic diagram for general configuration of IGCC installation, integrating the gasifier and syngas cleaning units with gas and steam turbines

were compared with two conventional IGCC with and without CO_2 capture. In the new IGCC concepts different options were considered, for instance, air-blown regenerator or oxygen-blown, cryogenic air separation unit for O_2 supply, or membrane separation, for flue gas treatment were also considered different choices, triple-pressure or single-pressure drum-type heat recovery steam generator for heat recovery or amine scrubber, *etc.* According to the authors, the new IGCC plants with CO_2 capture have efficiencies 18.5–21% higher than that of the conventional IGCC CO_2 capture plant. The CO_2 capture efficiencies of the new concepts were 95.8–97%, against 87.7% of the conventional plant. The investments costs for the proposed new plants varied in the range 1207–1479 €/kW$_e$, depending on the concept considered, while the investments costs of the conventional plants were 1620 €/kW$_e$ or 1134 €/kW$_e$, respectively, for plants with and without CO_2 capture. Therefore, Mondol *et al.* [9] concluded that the new plants were more efficient, cleaner, and produce electricity at a cheaper price than the conventional ones.

IGCC energy conversion efficiency is usually around 40–45%, but several authors have studied different configurations with the aim of increasing these figures, using specific or dedicated software, such as: ASPEN or ECLIPSE. Some of these new concepts also include CCS installations.

Brown *et al.* [10] studied the thermo-economic aspects of the conversion of biomass into energy. Different options were studied for fluidized bed gasifier operation: atmospheric or pressurized air, oxygen, or steam. The gasifier was connected to an internal combustion engine combined cycle ICE-CC or to gas turbine combined cycle (GT-CC). The ICE-CC required a cold gas cleaning system with tar concentrations below $100 \, mg/Nm^3$ and particulates lower than $50 \, mg/Nm^3$, while GT-CC demanded tar concentrations and particulates minor than $5 \, mg/Nm^3$ and $30 \, mg/Nm^3$, respectively. According to the authors' simulation results, electricity conversion efficiencies were higher for ICE-CC, while GT-CC led to lower investment costs and to optimal capital costs and electricity generation costs. The best specific capital costs were calculated for steam gasification, followed by air gasification and then for oxygen gasification, always for GT-CC. In relation to annual electricity generation costs the differences among different oxidants were narrow; however, air gasification was optimal for GT-CC and steam gasification for ICE-CC. According to the authors, the models used still need some improvement, for instance the reaction model needs further calibration to take into account product formation at different pressures and in presence of other oxidants than air.

Chen *et al.* [11] also studied the implications of adding a CCS unit to an IGCC plant. The studied system integrated a quench gasifier with water gas shift reactor and a Selexol system to remove sulfur and CO_2. These authors analyzed the effect of different parameters on plant performance and cost, such as cost and quality of coal gasified and CO_2 removal efficiency. As coal quality increased, gasification efficiency, thermal efficiency and capital cost of power plant also increased. The cost of removing CO_2 decreased with the rise of CO_2 removal efficiency, being 90% the optimal value. When advanced technologies for oxygen production and for gas turbines were considered the efficiency increased and the cost of IGCC systems decreased both with and without CCS. The joining of advanced technologies with CCS led to an increase in capital cost, but due to the higher efficiency the estimation for the cost of electricity was lower than that of current plant.

Many other authors have studied different configurations for IGCC (Lee JJ *et al.* [12], Wu C *et al.* [13], Christou *et al.* [14], Sudiro *et al.* [15], and Franco *et al.* [16]). Some of them integrated advanced technologies on which there is not enough information and data, and therefore some of these studies present significant uncertainties in relation to efficiency and cost estimations.

7.3.3 *Hydrogen Production*

The growing interest in hydrogen utilization is mainly due to environmental concerns and the security of fossil fuels supplies. The main advantage of using hydrogen as an energy carrier is the environmental benefits over fossil fuels. However, currently, hydrogen is mainly produced through fossil fuels. In the near future it is expected to be able to produce hydrogen from biomass gasification in a clean and efficient way. Nevertheless, there are some technical problems that need to be solved before biomass gasification may become a feasible way of producing hydrogen. One of the challenges is to increase hydrogen content in syngas through

WGS reaction (7.8). The conversion of CO with steam to produce H_2 is usually improved by moderate temperature, by removing the hydrogen to shift the WGS equilibrium to the right, or by using excess steam and suitable catalysts.

The removal of CO from the syngas is crucial if the objective is to use the hydrogen in fuel cells, as this compound is poisonous for this kind of application. The main challenge in this field is the development of suitable catalysts for this reaction, in order to make the hydrogen production from biomass gasification economically and technically more attractive.

Extensive research on WGS catalysts has been reported in the literature. There are two main groups of catalysts to be used in WGS reaction: those applied at higher temperature (320–450°C) and others operating at lower temperatures (200–250°C). In the first group, iron-chromium based catalysts are mainly used and in the second copper-zinc based catalysts. The Fe–Cr formulation was reported in 1912 and since then studies with different promoters such as B, Cu, Ba, Pb, Hg, Ag, B, Ce, Zn, and Co have been reported by Zhang et al. [17]. Other metals such as Sn, Ce, Ru, and Rd (Basinska et al. [18]) and zeolites (Souza et al. [19]) have also been tested to increase the WGS reaction performance. Apart from this, gold based catalysts and platinum group metals have been proved to facilitate the WGS reaction, as reported by several authors, including Yeung and Tsang [20] and Andreeva et al. [21].

After the WGS reaction, the separation of CO_2 from the gases is needed to obtain high purity hydrogen. There are four major possible ways of accomplish this separation: chemical or physical absorption, pressure swing adsorption (PSA), cryogenic separation, and membrane separation. For a preliminary choice of the separation technology one should take into account the operating temperature and pressure ranges to be used, the syngas composition (CO_2, concentration and nature of other components present in the feed stream), and process cost. All of these technologies have limitations: PSA works at room temperature and high pressure and recovers less of the feed-stream hydrogen; cryogenics separation is normally used only in large-scale facilities, because it has a high capital cost; current polymer membrane separation systems have limited temperature tolerance and are also more vulnerable to chemical damage from aromatics compounds and H_2S.

CO_2 can be separated from H_2 by chemical or physical absorption using liquid solvents. The main disadvantage of the absorption process is that, in all cases, the solvent recycling is energy and capital demanding. Solvents frequently used in chemical absorption are alkanolamines such as monoethanolamine (MEA), diethanolamine (DEA), di-isopropanolamine (DIPA), methyldiethanolamine (MDEA), and diglycolamine (DGA) (McKee [22], Ebner and Ritter [23]). Ammonia and alkaline salt solutions are also used. Before CO_2 removal, the syngas has to be cooled and cleaned to remove particulates and other impurities. Then the clean gas passes through an absorption tower, where the absorption solution is placed. The separation occurs by a selective absorption of CO_2 by the solvent, which reacts chemically with CO_2 producing a weakly bound compound. After the absorption step, CO_2 is released in a stripper tower by reducing the pressure or raising the temperature to approximately 120°C.

In the physical absorption process the solvent only interacts physically with the dissolved gas and the relative absorption of CO_2 in solvent is favored over other components of the gas mixture. The most common physical solvents used are organic compounds with low surface tension, such as: methanol (Rectisol Process) and glycol ethers (Selexol Process), propylene carbonate (Fluor solvent process), and sulfolane. Also calcium oxide (CaO), sodium hydroxide (NaOH), and potassium hydroxide (KOH) are used as absorption solvents. This process can be used in various applications, but is considered the best choice for applications at higher pressure (*i.e.*, IGCC). The most used adsorbents include aluminosilicate zeolite molecular sieves, titanosilicate molecular sieves, and activated carbons (Ebner and Ritter [23]).

The PSA process is, currently, the most used. Other adsorption techniques for CO_2 capture have been developed, *e.g.*, temperature swing adsorption (TSA) and electrical swing adsorption (ESA); however they are not so popular (McKee [22]). A typical PSA unit consists of a series of containers, each holding the same type of adsorbing material. The PSA separation technology is a cycling process with two basic steps, the gas compound adsorption and the adsorption material regeneration. The utilization of at least two adsorbent vessels allows an almost continuous process of obtaining hydrogen. The gases are separated according to the characteristics of the molecular species and affinity for a specific adsorbent material (*e.g.*, zeolites and activated carbon). After the process is completed it swings to low pressure to desorb the adsorbent material. The adsorption material regeneration takes place after the adsorption bed reaches the end of its capacity to adsorb CO_2, then the feed gas is switched to a second adsorption bed, and the first one is regenerated by depressurizing the adsorbent bed. It is then ready for another cycle of producing high purity hydrogen. Also, the off going gas of the vessel being depressurized can be used to pressurize partially the second vessel. This procedure has the advantage of saving a significant amount of energy, so it is commonly used on an industrial scale. The CO_2 obtained after regeneration shall be compressed for transport and storage. The parameters that were shown to have more effect on the adsorption efficiency are temperature, partial pressures, surface forces, and adsorbent pore sizes. The two main advantage of PSA process is high purity hydrogen production (up to 99.999 vol.%) and the ability of removing CO and CO_2 to very low concentrations (0.1–10 ppmv).

The TSA process is very similar to PSA, the main difference being that the adsorbent regeneration occurs by raising the temperature. The ESA process uses as adsorbent a carbon-bonded activated carbon fiber. Adsorbent efficiency to adsorb CO_2 can vary depending on the pore volume and size and surface. This is a new material that is highly conductive, so the desorption of the adsorbed gases occurs rapidly by low-voltage electrical current with no variation in the system pressure and with a very small modification on the system temperature.

Cryogenic separation uses gas condensation as the separation principle. The CO_2 is physically separated from H_2 by condensation at cryogenic temperatures. The phase modification (gas to liquid or solid) is induced by compressing and cooling the gas mixture in a multiple stage process, which is more efficient if the gas mixture contains compounds with significant differences in boiling points (Shackley and Gough [24]). Usually this process is only used for gas mixtures with very high CO_2 yields, usually higher than 90% (Shackley and Gough [24]). The presence of impurities (SO_2 and NO_x) can complicate the separation, so a previous gas cleaning is essential. Furthermore, the gas has to be dried before being cooled down, because the presence of water complicates the process. The main disadvantage of the cryogenic separation is the high energy needed for gas cooling and pressurization that make this process very expensive. An advantage of this process is that the liquid CO_2 is easier to transport since it does not need compression.

Membrane technology is considered as an attractive way of separating CO_2 from gas streams. A membrane is a selective barrier between two phases, in which some components pass through it while others are retained. The highest developments in membrane technology occurred during the 1980s, due to the development of synthetic polymeric membranes (Basu *et al.* [25]). Currently, commercially membranes of a wide range of materials are available, metallic, ceramic or organic. Nowadays, membranes are still too expensive and energy demanding to be implemented on a large scale.

The high processing costs associated with the absorbent/adsorbent regeneration and phase exchange (gas to liquid) are eliminated in membrane separation processes, which present certain advantages over other separation methods, namely low maintenance, low energy requirements, and being environmental friendly (Basu *et al.* [25]). Other advantages include compactness, light weight, and modular design, allowing a multi-stage operation.

For a membrane to be suitable for H_2 removal it needs to have a high selectivity for H_2 and high permeability. The high permeability is necessary to produce a compact membrane facility, since many current systems require a large membrane area to achieve the desired gas stream purity and flow rate. So, new membrane types are essential to attain high permeability and selectivity, as well as long-term durability.

7.3.4 Fischer–Tropsch Synthesis

The process by which CO undergoes hydrogenation over iron, cobalt, or nickel catalysts at atmospheric pressure and temperatures of 180–250°C, leading to a mixture of linear and branched hydrocarbons and oxygenated products, is named the Fischer–Tropsch synthesis (FTS). The FTS provides alternative routes for the production of transportation fuels and petrochemical feedstock. This process can be designed to produce gasoline, diesel, and/or chemicals. In 2002 two Fischer–Tropsch synthesis (FTS) plants existed as commercial operations, *i.e.*, Sasol in South Africa and Shell in Malaysia, but interest in the FTS technology has increased due to decreasing oil reserves, the geographic location of these reserves, the demand for cleaner feedstock, and the reduction of CO_2 emissions. In 2005 the

American company Rentech explored four pilot installations and a semi-commercial facility for FTS. The companies Sinopec and Syntroleum announced their intention to start building, in 2007, two installations, one in China and the other in Papua New Guinea (Boerrigter [26]).

The FTS has been specially used with syntheses gas produced from coal gasification (Boerrigter [27]). More recently, the interest in using this technology combined with biomass gasification has increased dramatically, due to the decision of the European Commission to substitute 20% of conventional fuels by alternative fuels in the road transport sector by the year 2020 (EU [28]). The combination of biomass gasification with FTS is designated as biomass-to-liquids (BTL). Nowadays, the main objective of the research and development projects are the production of second generation bio-fuels by biomass and/or wastes gasification combined with FTS.

Syngas produced by gasification process needs to be cleaned and conditioned to meet the required specifications to be used in FTS, as mentioned in Table 7.2. The extensive cleaning is needed because the catalyst lifetime is greatly affected by the presence of trace pollutants, which can lead to changes in the physical and chemical properties of the catalysts. Conditioning is needed to adjust the H_2/CO ratio to approximately 2, due to the stoichiometry of the FTS reactions. This adjustment is performed by water gas shift reaction (7.8), followed by a CO_2 removal unit. In FTS CO reacts with H_2 to produce mainly linear paraffins and α-olefins by reactions (7.17) and (7.18), which are highly exothermic reactions:

$$\text{Paraffins formation: } (2n+1)H_2 + nCO \leftrightarrows C_nH_{2n+2} + nH_2O \qquad (7.17)$$

$$\text{Olefins formation: } 2nH_2 + nCO \leftrightarrows C_nH_{2n} + nH_2O \qquad (7.18)$$

The most frequent FTS catalysts used have metals from groups 8, 9, and 10 (formerly group VIII) like Fe, Co, and Ru. The iron based catalysts are the most used in FTS due to their lower cost in comparison with other active metals. They are normally used in FTS using syngas from coal (Wu *et al.* [13]) but are promising option for biomass conversion (Steen and Claeys [29]). To obtain highly active FTS catalysts, the promotion of the iron based catalyst is required (Steen and Claeys [29]). Several promoters have been tested, but the potassium appears to be the most cost effective promoter (Luo and Davis [30]). A lot of work has been done to develop new and more efficient iron catalysts (Wu *et al.* [13]).

Cobalt based catalysts present the highest activity and generate predominantly linear alkanes. Also, these catalysts are not inhibited by water, resulting in a higher productivity and high synthesis gas conversion (Borg *et al.* [31]). The main disadvantages of these catalysts are the low water gas shift activity and high cost of cobalt. Inorganic supports with high surface area (*e.g.*, silica, alumina) have been studied to increase the surface area of these catalysts (Borg *et al.* [31], Bao *et al.* [32], Song and Li [33], Storsæter *et al.* [34]), but alumina appears to be the best choice (Steen and Claeys [29]). The use of various support materials such as SiO_2, Al_2O_3, and TiO_2 have been patented.

Ruthenium based catalysts are the most active for FTS, but the high price of ruthenium eliminate its application on the industrial scale. At low operation temperatures and high pressures the Ruthenium based catalysts are selective towards high molecular waxes, but at relatively low pressures produce a large amount of methane.

Many comparative studies of different catalysts for FTS have been published, but the catalytic activity of each of these catalysts with respect to the FTS reaction is still controversial. Also, the particle size and dispersion of the catalyst has an important role in its selectivity and activity. The conversion of gas to hydrocarbons (Gas-To-Liquids route) is currently one of the most promising topics in the energy industry due to economic utilization of wastes to produce environmentally clean fuels, which can have many applications. The FTS technology allows the utilization of biomass and wastes as feedstock in the fuels market.

7.3.5 Synthesis of Methanol and Dimethyl Ether

The methanol and dimethyl ether (DME) syntheses have attracted great interest because of their potential to be used as fuels and as chemicals. Methanol can be used directly or blended with other petroleum products as a clean burning transportation fuel and is also an important chemical intermediate used to produce a large number of chemicals.

The most widely used feedstock for methanol and DME production is natural gas, but other feedstocks can be employed. Coal is increasingly being used in methanol production, via gasification and syngas production, especially in China. Moreover, well established technologies already available are being applied for methanol production using the syngas obtained during biomass gasification, due to the possibility of energy production and greenhouse gas emissions reduction. Many studies were performed in methanol synthesis from biomass-derived syngas (Kumabe et al. [35], Zhang et al. [17]). Several biomass-to-methanol demonstration projects have been developed, such as the Hynol project in the United States, the BAL-Fuels (Biomass-Derived Alcohols Fuels), BioMeet, and BLGMF projects in Sweden, and the BGMSS project in Japan (Zhang et al. [17]).

The Syngas produced needs to be extensively cleaned and conditioned before it may be used for methanol or DME synthesis, the required specifications being similar to those for FT synthesis (Table 7.2). Also, a syngas with H_2/CO ratio of approximately 2 is needed, so a WGS reaction is required. Syngas produced during biomass gasification has a different composition from that derived from natural gas or coal. It contains higher amounts of CO_2 and lower amounts of H_2, which results in a low H/C ratio and a high CO_2/CO ratio (Yin and Leung [36]). This higher percentage of CO_2 led to the idea of using CO_2, the most important greenhouse gas, as an alternative feedstock, replacing CO in the methanol production. In the methanol synthesis CO and H_2 react to produce methanol by reaction (7.19). Methanol can also be produced by CO_2 hydrogenation, according to reaction (7.20). The carbon source for methanol synthesis is still under debate (Liu *et al.* [37]):

$$CO + 2\,H_2 \rightleftarrows CH_3OH \qquad\qquad (7.19)$$

$$CO_2 + 2H_2 \rightleftarrows CH_3OH + H_2O \qquad\qquad (7.20)$$

The main challenge of methanol and DME synthesis is the development of more efficient catalysts. Several research groups have been working in catalyst preparation using different catalyst compositions and preparation methods; however, uncertainties still remain about the role of active catalyst sites, the effect of various promoters addition, and reaction mechanism. Recently, new catalysts based on nickel, copper, zinc and alloys, and also ultrafine particle catalysts have been proposed for methanol synthesis (Venugopal *et al.* [38], Kiss *et al.* [39]).

The properties of catalysts used in methanol synthesis have been extensively studied, but Cu still continues to be an important active catalyst component (Liu *et al.* [37]). Also, some studies on catalysts for methanol synthesis from CO_2 hydrogenation have been performed (Liu *et al.* [37], Liang *et al.* [40]). These studies showed that the Cu/ZnO-based catalysts modified with different metals or oxides exhibit considerable activity and selectivity for the methanol production via CO_2 hydrogenation (Melián-Cabrera *et al.* [41], Liu *et al.* [37], Liang *et al.* [40]).

The DME synthesis consists of two steps: the methanol synthesis followed by methanol dehydration by reaction (7.21):

$$2CH_3OH \rightleftarrows CH_3OCH_3 + H_2O \qquad\qquad (7.21)$$

The research studies have been focused on the development of better catalysts with higher selectivity for DME formation. Some research groups tried to find bifunctional catalysts (Ge *et al.* [27]). These catalysts present two types of active sites: one for methanol synthesis and the other for methanol dehydration (Ge *et al.* [27]). Others have reported that the catalysts more suitable for DME synthesis were mostly Cu/ZnO based catalysts (methanol synthesis catalyst) mixed with a solid acid catalyst, such as γ-alumina or zeolites (methanol dehydration catalysts) (Yaripour *et al.* [42], Venugopal *et al.* [38]).

The key issue to chemical synthesis from syngas is the development of more efficient and lower cost catalysts. Many studies have been performed, but there are still a lot of improvements to be achieved.

7.4 The Role of Gasification in CCS and in Global Warming Abatement

Many countries compromised to decrease greenhouse gas emissions since the Kyoto protocol. To achieve this goal several strategies may be followed, increasing electricity and power generation efficiency, raising the role of bio-wastes for energy production, and increasing the share of renewable and nuclear sources for energy production. However, nuclear energy is not well accepted by common citizens and most renewable technologies are not yet sufficiently advanced to allow fossil fuels substitution. Therefore, according to current predictions, fossil fuels like coal, oil, and natural gas will continue to have the greatest contribution for energy production, at least till 2030, around 85% of today's needs, IEA [2]. Consequently, it is of great importance to develop new technologies for energy production from fossil fuels that could diminish CO_2 emissions, facilitate CO_2 capture and storage, and increase energy efficiency.

The use of coal for electricity production will continue to be significant and technologies for clean coal are most needed. Different options may be considered for CO_2 capture, as shown in Figure 7.5:

- pre-combustion, (production of syngas by gasification or pyrolysis processes, conversion of CO into CO_2, and CO_2 capture);
- oxy-combustion (combustion with pure O_2 with recycled flue gas and purification of CO_2 to remove impurities and incondensable gases);
- post-combustion (air combustion and removal of CO_2 from exhaust gases).

In coal combustion systems with air, CO_2 is emitted in large quantities and its sequestration is not very attractive, because CO_2 is diluted in N_2 and large amounts of flue gases need to be treated for N_2 separation from CO_2, prior to CO_2 sequestration by expensive processes. In fact, CO_2 adsorption, membranes, and cryogenic separation are not suitable. Cryogenic separation needs much energy, due to the low content of CO_2 in the exhaust gases and is too expensive, especially for gases at atmospheric pressure. Membranes are not suitable due to the existence of dust, SO_x, NO_x and incondensable gases, and due to membranes physical degradation. CO_2 adsorption on a solid is also not adequate because of exhaust gases high flows and impurities. Chemical absorption technologies are probably the most adequate. However, the choice of the method depends on exhaust gas characteristics and, though different amines have been used for this purpose, they are degraded by common impurities of exhaust gases. Therefore, the main challenges for chemical absorption processes are resistance to degradation caused by exhaust gases impurities and the need for a high capacity of regeneration.

Therefore, conventional pulverized fuel systems and circulating fluidized beds are being converted to oxy-combustion, in which O_2 mixed with recirculated flue gases is used instead of air. Thus, a flue gas stream with high concentrations of CO_2 and mainly containing CO_2 and H_2O is produced, which makes CO_2 separation easier. However, purification of the CO_2 flow to remove incondensable gases is still needed.

Besides this oxy-combustion process (O_2/CO_2 recycle) followed by post-combustion capture, another option for CCS is pre-combustion capture, in which fuel carbon content is removed before combustion and a CO_2 by-product stream is produced, together with a hydrogen rich fuel. Therefore, coal is first gasified to produce syngas, whose main components are CO and H_2. Syngas may also be produced from natural gas by steam reforming or partial oxidation. Next, CO is

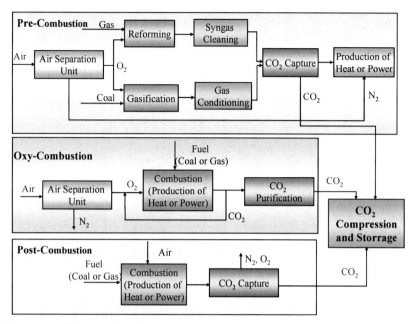

Figure 7.5 Different options for CO_2 capture

converted into more H_2 and CO_2 in the presence of steam by water-shift reaction. Therefore, the final CO_2 concentration is much higher, with an easier CO_2 separation process. Different technologies for CO_2 separation are under development, the most promising being pressure swing adsorption (PSA) and membrane or cryogenic separation, as mentioned in Section 7.3.3. These technologies are more attractive for syngas than for exhaust gas, because syngas is cleaner and, due to its higher pressure and high CO_2 content, it is possible to liquefy it by cooling. Afterwards, the produced CO_2 rich stream is going to storage and the H_2-rich fuel can be used for energy production in a gas turbine combined cycle or in fuel cells to produce electricity.

Gnanapragasam *et al.* [43] studied the effect of gasification operating conditions on reducing CO_2 emissions for an IGCC power generation system. As reported by these authors, the release of CO_2 depends on gasification conditions, namely higher O_2 contents leads to the formation of higher CO_2 amounts, whilst the rise of steam input decreases the release of CO_2. The type of fuel gasified also affects the formation of CO_2. Gnanapragasam *et al.* [43] studied four different types of solid fuels including coal and biomass species and verified that the highest CO_2 contents were obtained with wood chips. CO_2 formation is also affected by the temperature conditions of the IGCC unit; to decrease CO_2 it is important not to increase the compressed air temperature prior to its entrance in the gas turbine combustion chamber. The lowest CO_2 emissions were also obtained for a lower inlet temperature in the heat recovery steam generator and a higher gas turbine inlet temperature.

As analyzed in Section 7.3.2, electricity cost is higher when CO_2 capture units are included, because CO_2 removal needs a great deal of energy, thus decreasing process global efficiency. Kanniche *et al.* [44] calculated the cost of electricity production from coal and gas when CO_2 is captured and also the cost of each CO_2 ton captured. These authors studied different options for CO_2 capture: pre-combustion, oxy-combustion and post-combustion. For the first option IGCC was analyzed, considering two gasification types, a conventional process with gasification of dry coal, and with classical combined cycle, producing a gross power output of 320 MWe and a new technology with coal and water slurry gasification integrated in a advanced combined cycle (with steam cooling of the combustion turbine blades), producing a gross power output of 1200 MWe.

For the oxy-combustion option, two types of pulverized coal (PC) power stations were analyzed by Kanniche *et al.* [44]: a sub-critical power station, whose gross power output was 600 MWe and a super-critical power station with a gross power output of 1200 MWe. Two NGCC (Natural Gas Combined Cycle) combined cycles were also studied, one with a 9H type combustion turbine, an evaporation boiler line and a single shaft steam turbine, supplying a gross power of 480 MWe and the other with 9H type combustion turbines, two evaporation boilers and one steam turbine line, which provided 960 MWe. NGCC could be modified to the three capture methods: pre-combustion, oxy-combustion, and post-combustion. In the pre-combustion option, methane was reformed, while CO was converted into CO_2, which is afterwards captured. According to Kanniche *et al.* [44] this option is more expensive than the others, being only attractive for hydrogen production.

Kanniche *et al.* [44] results showed that the highest efficiency was obtained for NGCC with post-combustion capture (50%), being followed by oxy-combustion in PC (35%), and by IGCC (33.5%), and the lowest efficiency was obtained for post combustion capture in PC (30%). In relation to investment the least expensive technology was NGCC, followed by PC and IGCC with slurry. Oxy-combustion PC and IGCC with slurry led to the lowest production costs. IGCC with slurry, together with oxy-combustion in PC and the current IGCC led to the lowest costs per ton of CO_2 capture. The highest value was obtained for NGCC pre-combustion capture. With the results obtained it was difficult to select the best option for CO_2 capture.

Another method for CO_2 capture is chemical looping (CLC), which consists of two reactors. An oxygen carrier metal is used inside the gasifier which, by being reduced, supplies the oxygen needed for fuel gasification to produce syngas. The reduced metal oxide goes to another reactor, where it is again oxidized in the presence of air. Different metals may be used as oxygen carriers, such as nickel, manganese, calcium, or iron oxides stabilized in a support material like alumina or zirconia. CaO is the most used oxygen carrier, which inside the gasifier is converted into $CaCO_3$. The solid produced in the gasifier also contains CaS, char, and ash, which goes into the regenerator, where $CaCO_3$ is again transformed into CaO, producing a CO_2 stream with high purity ready for storage, as shown in Figure 7.6.

This technology does not need sulfur removal units, water gas shift reactors, or membranes, as all these processes are included in reaction (7.22). Char oxidation

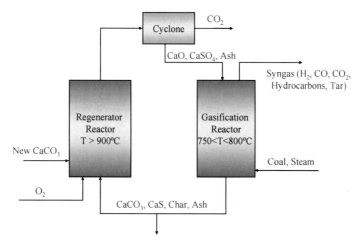

Figure 7.6 Schematic diagram of conventional chemical looping

in the regeneration reactor supplies some of the heat needed for the regeneration reactor, reaction (7.23). The use of coal with high ash and sulfur content oblige to frequent solids purge:

$$Coal + aCaO + H_2O \leftrightarrows H_2 + (a-z) CaCO_3 + yC + zCaS \qquad (7.22)$$

$$(a-z)CaCO_3 + yC + zCaS + bO_2 \leftrightarrows (a-z)CaO + zCaSO_4 + (a-z-y)CO_2 \qquad (7.23)$$

Rezvani *et al.* [45] analyzed the techno-economic viability of different CO_2 capture technologies, using the simulation software ECLIPSE: physical absorption, water gas shift reactor membranes, and two chemical looping combustion cycles (CLC) with single and double stage reactors.

A water gas shift reactor was used to convert CO into CO_2, which was removed by physical absorption processes with Selexol solvents and is then compressed to 110 bar for pipeline transportation. The H_2 produced was mixed with N_2 and went into a gas turbine for power generation. The exhaust gases went to a steam generator and next to a steam turbine. In another configuration a water gas shift membrane reactor (WGSMR) and an oxygen transport membrane (OTM) were considered to increase power plant efficiency. OTM was used to recover the remaining gas combustibles in the retentate side of WGSMR.

In the CLC option the gas that left the oxidation reactor went to a gas turbine for power generation and then went through a steam generator to produce steam for a steam turbine for more power generation. The flue gas that left the fuel reactor also went to a gas turbine and next to a steam generator and through the gas condenser prior to the CO_2 compression unit. Because of the exothermic reactions, temperature may reach values that lead to sintering and agglomeration of the oxygen carrier material. To reduce the temperature, the oxygen carrier metal may be cooled down with additional air or, alternatively, by a double stage CLC reactor. In this new configuration, the air that left the oxidation reactor went into a gas turbine, where both temperature and pressure decreases before it went into the second oxidation reactor. Associated with each oxidation reactor there was a fuel reactor, the flue gases from these two reactors going to the same gas turbine, while there was a separate gas turbine for each exhaust gas coming from the oxidation reactor. The gases leaving each gas turbine were used to produce steam for steam turbines to produce electricity.

Rezvani *et al.* [45] studies showed that the membrane option was a promising one, though the development of hydrogen selective membranes that are more economic and efficient is still needed. The CLC options had high costs, were not able to produce H_2, and further research is needed before a robust technology is available.

CCS still has many doubts, uncertainties and knowledge gaps, especially concerning life-cycle effects, storage capacity, and permanence and cost. However, these difficulties must not be taken as an excuse for stopping research studies. More technical and engineering data are still needed since not all the technologies analyzed are at the same stage of development and deployment and some of them still need further demonstration on a larger scale. On the other hand, economic estimations are dependent on fuels and other materials supply restrictions, and on equipment costs, which may suffer alterations depending on the maturity of technologies and on the amount of equipment produced. Therefore more evaluation studies are still needed for accurate technical, economic, and environmental estimations.

Though CO_2 capture is expensive and more fuel is spent to produce the same amount of electricity, effective CCS is fundamental for sustainable development, to prevent climate change, and to guarantee that life on planet Earth will continue as we know it. To achieve these goals and to accomplish zero emissions, biomass and wastes gasification will probably have an important role. The success of CCS also depends on strong policy framework, and governments of all nations, especially those from G8, have an important role in establishing sufficient and long-term incentives for CCS and for building CO_2 transportation networks.

7.5   Main Gasification Barriers and R&D Needs

Through gasification technologies it is possible to produce heat, power, CHP (combined heat and power), and synthesis gas. However, the main challenge for gasification technologies is the accomplishment of higher efficiency energy conversion, low environmental impact, and low cost in order to ensure sustainable development. It is desirable that gasification plays an important role in energy production in the near future, as it permits an easier CO_2 separation than other technologies. However, there are still some barriers that gasification needs to overcome:

- flexibility of gasification systems to diversify the type of feedstock: fossil fuels, different types of biomass and wastes or blends of both;
- efficient syngas cleaning processes;
- high efficiency of integrated gasification combined cycle (IGCC);

- development of efficient syngas chemical synthesis;
- process scale-up and fabrication;
- process demonstration at commercial scale;
- sustainable policies.

Gasification of low grade coals and of different types of wastes with high contents of undesirable elements may lead to the release of pollutants precursors into the syngas. Some of these feedstocks may also be more difficult to gasify, thus requiring more severe gasification conditions, new materials and/or coatings for existing materials, and the use of more expensive catalysts or sorbents to guarantee syngas quality. One of the key issues for gasification technologies spread is the development of effective gas cleaning processes to ensure the production of high quality syngas even when poor or low quality feedstocks are gasified. Research and development of low cost and efficient catalysts is most needed to achieve effective abatement of undesirable syngas components, like tar, S, N, and halogens compounds. Catalyst regeneration and the development of multi-function catalysts for simultaneously reducing different types of compounds, thus eliminating some steps of syngas cleaning treatments, are important issues. Technical, economical, and environmental viability improvements of the overall process are dependent on new catalysts development.

The success of liquid fuels production from syngas through chemical synthesis is also dependent on advances on more selective catalysts, for the production of specific compounds, thus allowing the simplification of products separation processes and the improvement of technical and economical viability of the chemical synthesis process. The key issue in pre-combustion technologies for CCS is CO_2 separation, mainly for economic reasons; thus further R&D is still needed to raise process efficiency and decrease materials costs, such as high performance membranes. Long term testing on a commercial scale of new technologies such as IGCC, chemical synthesis and CCS is fundamental to guarantee reliable operation and to validate these processes. On the other hand, as IGCC, chemical synthesis, and CCS installations spread, equipment and materials will be produced on a larger scale, which will decrease production costs.

All the R&D and demonstration activities still need political and funding support through research financing programs, adequate incentives, policies, and strategies. Without public commitment and governmental support CCS and energy production through zero emission technologies will not be accomplished and life on Earth will be put at serious risk.

References

1. COM (2008) 19 Promotion of the use of energy from renewable sources. Commission of the European Communities, Brussels; 23 January 2008
2. IEA (2008) World Energy Outlook 2008. International Energy Agency OECD/IEA Paris

3. Pinto F, Lopes H, André RN et al. (2007) Effect of catalysts in the quality of syngas and by-products obtained by co-gasification of coal and wastes. 1. Tars and nitrogen compounds abatement. Fuel 86:2052–2063

4. Pinto F, André RN, Franco C et al. (2009) Co-Gasification of coal and wastes in a pilot-scale installation: 1. Effect of catalysts in syngas treatment to achieve tars abatement. Fuel doi:10.1016/j.fuel.2008.12.012

5. Pinto F, Lopes H, André RN et al. (2008) Effect of catalysts in the quality of syngas and by-products obtained by co-gasification of coal and wastes 2: Heavy metals, sulphur and halogen compounds abatement. Fuel 87:1050–1062

6. Ohtsuka Y, Tsubouchi N, Kikuchi T et al. (2009) Recent progress in Japan on hot gas cleanup of hydrogen chloride, hydrogen sulfide and ammonia in coal-derived fuel gas. Powder Technol 190:340–347

7. Sridhar G, Paul PJ, Mukunda HS (2001) Biomass derived producer gas as a reciprocating engine fuel-an experimental analysis. Biomass Bioenerg 21:61–72

8. Frey HC, Zhu Y (2006) Improved system integration for integrated gasification combined cycle (IGCC) systems. Environ Sci Technol 40:1693–1699

9. Mondol JD, McIlveen-Wright D, Rezvani S et al. (2009) Techno-economic evaluation of advanced IGCC lignite coal fuelled power plants with CO_2 capture. Fuel doi:10.1016/j.fuel.2009.04.019

10. Brown D, Gassner M, Fuchino T et al. (2009) Thermo-economic analysis for the optimal conceptual design of biomass gasification energy conversion systems. Appl Therm Eng 29:2137–2152

11. Chen C, Rubin ES (2009) CO_2 control technology effects on IGCC plant performance and cost. Energ Policy 37:915–924

12. Lee JJ, Kim YS, Cha KS et al. (2009) Influence of system integration options on the performance of an integrated gasification combined cycle power plant. Appl Energ 86:1788–1796

13. Wu C, Yin X, Ma L et al. (2008) Design and operation of a 5.5 MWe biomass integrated gasification combined cycle demonstration plant. Energ Fuel 22:4259–4264

14. Christou C, Hadjipaschalis I, Poullikkas A (2008) Assessment of integrated gasification combined cycle technology competitiveness. Renew Sust Energ Rev 12:2459–2471

15. Sudiro M, Bertucco A, Ruggeri F et al. (2008) Improving process performances in coal gasification for power and synfuel production. Energ Fuel 22:3894–3901

16. Franco A, Diaz AR (2008) The future challenges for "clean coal technologies": joining efficiency increase and pollutant mission control. Energ doi:10.1016/j.energy.2008.09.012

17. Zhang L, Millet J-MM, Ozkan US (2009) Effect of Cu loading on the catalytic performance of Fe–Al–Cu for water-gas shift reaction. Appl Catal A-Gen 357:66–72

18. Basinska A, Klimkiewicz R, Teterycz H (2004) Catalysts of alcohol condensation tested in water gas shift reaction. React Kinet Catal Lett 82:271–277

19. Souza TRO, Brito SMO, Andrade HMC (1999) Zeolite catalysts for the water gas shift reaction. Appl Catal A Gen 178:7–15

20. Yeung CMY, Tsang SC (2009) Noble metal core-ceria shell catalysts for water-gas shift reaction. J Phys Chem C 113:6074–6087

21. Andreeva D, Ivanov I, Ilieva L et al. (2009) Gold catalysts supported on ceria doped by rare earth metals for water gas shift reaction: Influence of the preparation method. Appl Catal A Gen 357:159–169

22. McKee B (2002) Solutions for the 21st century, zero emissions technologies for fossil fuels. International Energy Agency, Technology Status Report

23. Ebner AD, Ritter JA (2009) State-of-the-art adsorption and membrane separation processes for carbon dioxide production from carbon dioxide emitting industries. Separ Sci Technol 44:1273–1421

24. Shackley S, Gough C (2006) Carbon capture and its storage: an integrated assessment. Ashgate Publishing Ltd. Edition

25. Basu A, Akhtar J, Rahman MH *et al.* (2004) A review of separation of gases using membrane systems. Pet Sci Technol 22:1343–1368
26. Boerrigter H (2006) Economy of biomass-to-liquid (BTL) plants. Energy research Centre of the Netherlands (ECN), ECN-C-06-019
27. Ge Q, Huang Y, Qiu F *et al.* (1998) Bifunctional catalysts for conversion of synthesis gas to dimethyl ether. Appl Catal A Gen 167:23–30
28. European Union (2003) The promotion of the use of biofuels or other renewable fuels for transport. Directive EU 2001/0265(COD)
29. Steen E, Claeys M (2008) Fischer–Tropsch catalysts for the biomass-to-liquid process. Chem Eng Technol 31:655–666
30. Luo M, Davis BH (2003) Fischer–Tropsch synthesis: group II alkali-earth metal promoted catalysts. Appl Catal A-Gen 246:171–181
31. Borg Ø, Eri S, Blekkan EA *et al.* (2007) Fischer–Tropsch synthesis over γ-alumina-supported cobalt catalysts: effect of support variables. J Catal 248:89–100
32. Bao A, Liew K, Li J (2009) Fischer–Tropsch synthesis on CaO-promoted Co/Al$_2$O$_3$ catalysts. J Mol Catal A Chem 304:47–51
33. Song D, Li J (2006) Effect of catalyst pore size on the catalytic performance of silica supported cobalt Fischer–Tropsch catalysts. J Mol Catal A Chem 247:206–212
34. Storsæter S, Borg Ø, Blekka EA *et al.* (2005) Study of the effect of water on Fischer–Tropsch synthesis over supported cobalt catalysts. J Catal 231:405–419
35. Kumabe K, Hanaoka T, Fujimoto S *et al.* (2006) Co-gasification of woody biomass and coal with air and steam. Fuel, 86:684–689
36. Yin X, Leung DYC (2005) Characteristics of the synthesis of methanol using biomass-derived syngas. Energ Fuel 19:305–310
37. Liu X-M, Lu GQ, Yan Z-F *et al.* (2003) Recent advances in catalysts for methanol synthesis via hydrogenation of CO and CO$_2$. Ind Eng Chem Res 42:6518–6530
38. Venugopal A, Palgunadi J, Deog JK *et al.* (2009) Dimethyl ether synthesis on the admixed catalysts of Cu-Zn-Al-M (M = Ga, La, Y, Zr) and γ-Al$_2$O$_3$: the role of modifier. J Mol Catal A-Chem 302:20–27
39. Kiss J, Witt A, Meyer B *et al.* (2009) Methanol synthesis on ZnO(0001). I. Hydrogen coverage, charge state of oxygen vacancies, and chemical reactivity. J Chem Phys 130:184706
40. Liang X-L, Dong X, Lin G-D *et al.* (2009) Carbon nanotube-supported Pd–ZnO catalyst for hydrogenation of CO$_2$ to methanol. Appl Catal B-Environ 88:315–322
41. Melián-Cabrera I, López GM, Fierro JLG (2003) Pd-modified Cu–Zn catalysts for methanol synthesis from CO$_2$/H$_2$ mixtures: catalytic structures and performance. J Catal 210:285–294
42. Yaripour F, Mollavali M, Jam SM *et al.* (2009) Catalytic dehydration of methanol to Dimethyl ether catalyzed by aluminum phosphate catalysts. Energ Fuel 23:1896–1900
43. Gnanapragasam N, Reddy B, Rosen M (2009) Reducing CO$_2$ emissions for an IGCC power generation system: effect of variations in gasifier and system operating conditions. Energy Convers Manage doi:10.1016/j.enconman.2009.04.017
44. Kanniche M, Gros-Bonnivard R, Jaud P *et al.* (2009) Pre-combustion, post-combustion and oxy-combustion in thermal power plant for CO$_2$ capture. Appl Therm Eng, doi: 10.1016/j.applthermaleng.2009.05.005
45. Rezvani S, Huang Y, McIlveen-Wright D *et al.* (2009) Comparative assessment of coal fired IGCC systems with CO2 capture using physical absorption, membrane reactors and chemical looping. Fuel doi:10.1016/j.fuel.2009.04.021

Chapter 8
The Integration of Micro-CHP and Biofuels for Decentralized CHP Applications

Aggelos Doukelis and Emmanouil Kakaras

Abstract Renewable micro-CHP systems are a combination of micro-CHP technology and renewable energy technology, such as biomass gasification systems or solar concentrators. The integration of renewable energy sources with micro-CHP allows for the development of sustainable energy systems with the potential for high market penetration; a cost-effective and reliable heat and electricity supply; and a highly beneficial environmental and economical impact on a pan-European scale. The purpose of this chapter is to present results from the European co-ordination action project MICROCHEAP that intended to bring together industrial specialists and research experts to focus entirely on renewable micro-CHP technology, co-ordinate and steer research in this field, and highlight the most promising technologies with the highest potential for market penetration in existing and future market conditions. The chapter discusses the state of the art technological options in the field of renewable micro-CHP with biofuels with regards to technology, cost, and environmental impacts and presents a market survey concerning the possibility of future penetration of the technology in Europe. The results will provide a coherent overview of the basic technological options for renewable micro-CHP with biofuels and will provide an insight to the market trends within Europe and projected future market scenarios, taking into account cost estimations for various micro-CHP technologies, feedstocks, and electricity and fuel prices in Europe.

A. Doukelis (✉)
National Technical University of Athens, School of Mechanical Engineering,
Lab. of Steam Boilers and Thermal Plants,
Heroon Polytechniou 9, Athens 15780, Greece
Tel: +30 210 7722720, Fax: +30 210 7723663
e-mail: adoukel@central.ntua.gr

8.1 Introduction

Conventional power generation consists of large central plants connected to a grid supplying power to the customer. While such systems offer centralization they suffer from significant thermal losses as well as line transmission losses. Critical is the distribution grid that must be maintained and expanded to meet growth. To expand the grid realistically in today's environment is difficult. Not only from a practical perspective but unattainable in most metropolitan areas simply due to density and the NIMBY principal (Not In My Back Yard). The end result is the customer pays for these inefficiencies not only in cost of power but also in power outages, power sags and surges, and unstable frequency, *etc.*

Combined heat and power (CHP) systems (also known as cogeneration) generate electricity and thermal energy in a single, integrated system. In a traditional power plant that delivers electricity to consumers, about 30% of the heat content of the primary heat energy source reaches the consumer, although the efficiency can be 20% for very old plants and 45% for newer gas plants. In contrast, a CHP system converts 15–42% of the primary heat to electricity, and most of the remaining heat is captured for hot water or space heating. In total, as much as 90% of the heat from the primary energy source goes to useful purposes when heat production does not exceed the demand. Cogeneration works in parallel with the utilities grid reducing the strain. In critical power quality needs it can act as a filter to bring true high quality power that meets the client's needs. In some cases (depending on local and state regulation) a cogeneration plant can be designed to generate excess power for sale onto the grid creating an additional revenue source for the client.

CHP systems have benefited the industrial sector since the energy crisis of the 1970s. For three decades, these larger CHP systems were more economically justifiable than micro-CHP, due to the economy of scale. After the year 2000, micro-CHP has become cost effective in many markets around the world, due to rising energy costs. The development of micro-CHP systems has also been facilitated by recent technological advances in small heat engines. This includes improved performance and/or cost-effectiveness of {Stirling, steam, diesel, Otto} engines and gas turbines. The main difference of micro-CHP systems from their larger-scale kin is in the operating parameter-driven operation. In many cases industrial CHP systems primarily generate electricity and heat is a useful by-product. In contrast, micro-CHP systems, which operate in homes or small commercial buildings, are driven by heat-demand, delivering electricity as the by-product. Because of this operating model and because of the fluctuating electrical demand of the structures in which they would tend to operate, homes and small commercial buildings, micro-CHP systems will often generate more electricity than is instantly being demanded.

Renewable micro-CHP systems, on the other hand, are a combination of micro-CHP technology and renewable energy technology, such as biomass gasification systems or solar concentrators. The integration of renewable energy sources with micro-CHP allows for the development of sustainable energy systems with the potential for high market penetration. These technologies potentially have much to offer in helping us to achieve our objectives of tackling climate change, ensuring cost-effective and reliable heat and electricity supply and tackling fuel poverty. As well as providing low carbon energy to homes and small commercial buildings, micro-generation can provide the same service to community buildings, such as leisure centers and schools. In such premises, not only does the micro-generation installation help to reduce carbon emissions; it can also help to educate and inform communities about energy and, hopefully, persuade people to reduce their own carbon footprint.

The current work presents results from the European co-ordination action project MICROCHEAP that intended to bring together industrial specialists and research experts to focus entirely on renewable micro-CHP technology, co-ordinate and steer research in this field, and highlight the most promising technologies with the highest potential for market penetration in existing and future market conditions. The chapter discusses the state of the art technological options in the field of renewable micro-CHP with biofuels with regards to technology, cost, and environmental impacts, and presents a market survey concerning the possibility of future penetration of the technology in Europe.

8.2 State of the Art of Micro-CHP with Biofuels

Directive 2004/8/EC of 11 February 2004 "on the promotion of co-generation based on a useful heat demand in the internal energy market" provides the following definition of micro-CHP: 'micro-cogeneration unit' shall mean a cogeneration unit with a maximum capacity below 50 kWe. The suite of technologies caught by this definition includes solar (photovoltaics – PV – to provide electricity and thermal to provide hot water), micro-wind (including the new rooftop mounted turbines), micro-hydro, heat pumps, biomass, micro combined heat and power (micro CHP), and small-scale fuel cells [1]. The current work will focus on micro-CHP technologies that can be used in combination with biofuels, and will focus on three main systems using the most commercially available to date technologies: micro-turbines, Stirling engines, and fuel cells.

8.2.1 Micro-Turbines for Micro-CHP

The operating theory of the micro-turbine is similar to the gas turbine, except that most designs incorporate a recuperator to recover part of the exhaust heat for preheating the combustion air and increase the electric efficiency. Air is drawn through a compressor section, mixed with fuel, and ignited to power the turbine section and the generator. Hot exhaust gas from the turbine section is available for

Table 8.1 Current manufacturers of micro-turbines (less than 200 kWe)

Manufacturer	Range of models
Capstone Turbine Corporation	30, 60 kWe (next 200 kWe)
Elliot Energy Systems Inc.	80 kWe
Turbec AB	100 kWe
Bowman Power Ltd.	50, 80 kWe
Ingersoll–Rand Energy Systems	70 kWe (next 250 kWe)

CHP applications (hot water heating or low pressure steam applications). The high frequency power that is generated is converted to a grid compatible 50 Hz through power conditioning electronics. Their compact and lightweight design makes micro-turbines an attractive option for many light commercial/industrial applications.

Most manufacturers are pursuing a single-shaft design in the 25–250 kW range, where the compressor, turbine, and permanent-magnet generator are mounted on a single shaft supported on lubrication-free air bearings. These turbines operate at speeds of up to 120,000 rpm and can be operated on a variety of fuels, such as natural gas, gasoline, diesel, alcohol, biofuels, *etc.* With recuperation, efficiency is currently in the 25–30% LHV range, or even higher for new designs. Table 8.1 shows a list of the current manufacturers of micro-turbines below 200 kWe [2], while Table 8.2 presents the general specifications of market micro-turbine cogeneration systems [3].

Installed prices of $800–1200/kW for CHP applications is estimated when micro-turbines are mass produced, while availability is similar to other competing distributed resource technologies, *i.e.*, in the 90–95% range. Micro-turbines have substantially fewer moving parts than engines. The single shaft design with air bearings will not require lubricating oil or water, so maintenance costs should be below conventional gas turbines. Micro-turbines that use lubricating oil should not require frequent oil changes since the oil is isolated from combustion products. Only an annual scheduled maintenance interval is planned for micro-turbines.

Table 8.2 General specifications of some market micro-turbine CHP systems [3]

	Capstone micro-turbine					Elliot/ Bowman	Turbec
	NG/ gaseous propane	Diesel or kerosene	Biogas (landfill or digester gas)	NG		NG, propane, LPG, and butane	NG
Electrical capacity (kWel)	30	30	30	28	60	80	105
Electrical efficiency (%) LHV	26	25	26	25	28	28	30
Overall efficiency (%) LHV	91	90	91	91	89	75	78
Thermal output (kW)	85	85	85	85	150	136	167

Table 8.3 Micro-turbine emission characteristics [3]

	Capstone model 330	Ingersoll–Rand Energy Systems 70LM (two shaft)	Turbec T100
Nominal capacity, kW$_{el}$	30	70	100
Electrical efficiency (%) HHV	23	25	27
NOx, ppmv	9	9	15
CO, ppmv	40	9	15
THC, ppmv	< 9	< 9	< 10
CO$_2$, lb/MWh	1928	1774	1706
Carbon, lb/MWh	526	484	465

Maintenance costs are being estimated at 0.006–0.01\$/kW. Micro-turbines also promise lower noise levels and can be located adjacent to occupied areas [4–6].

Micro-turbines have the potential for producing low emissions. They are designed to achieve low emissions at full load; however, emissions are higher when operating under reduced load. Today's micro-turbines have a greater efficiency and lower emissions of greenhouse gases than internal combustion engines [7]. Low emission combustion systems are being demonstrated that provide emissions performance comparable to larger CHP turbines. The main pollutants from the use of micro-turbine systems are NO$_x$, CO, CO$_2$, and unburnt hydrocarbons, and negligible amount of SO$_2$. NO$_x$ emissions are targeted below 9 ppm using lean pre-mix technology without any post combustion treatment. Emission characteristics of micro-turbine systems based on manufacturers' guaranteed levels are given in Table 8.3.

8.2.2 Stirling Engines for Micro-CHP

Stirling engines are closed cycle engines operating on the Stirling cycle, characterized by an external heat supply. An external heat supply allows the use of any heat source operating at a sufficient temperature level. The Stirling engine operates by continuous heating and cooling of a fully enclosed working gas, usually helium, air/nitrogen or hydrogen. The alternate compression and expanding of a fixed amount of high pressure gas is transformed into a rotating movement to which the electric generators are connected. The continuous external combustion process of the Stirling engine provides good combustion control and low exhaust emission levels. Within the closed Stirling cycle, pressure variations of the working gas follow an almost sinusoidal curve, which is one of the basic reasons for the low noise and vibration level of a Stirling engine. Another reason for the low noise is that there is no connection between the working gas and the outside atmosphere as opposed to the exhaust pipe of an IC engine.

The Stirling engine was first patented in 1816 by Robert Stirling as an application of the regenerator he had invented. Throughout the nineteenth century, several working models were built and some were operated successfully for a while, but eventually more effective and powerful steam and internal combustion engines replaced the Stirling engines. The development of the modern high speed Stirling engine with a pressurized gas cycle started in the 1930s. The Philips electric company did pioneering work and although Philips abandoned Stirling development in the 1970s, most Stirling engines that are on the market today use solutions that were originally developed at Philips.

Two main families represent modern Stirling engines: kinematic and free piston Stirling engines. Kinematic Stirling engines are the largest group of Stirling engines. In these engines, the reciprocating movement of the power piston is transferred to a rotating shaft by mechanical means. The engines can be used to drive an electric generator. Depending on the geometric arrangement of the pistons (and displacers), different variants are the alpha, beta, or gamma engines. Depending on the interconnection of multiple cylinders, these engines can operate with single or double acting pistons. Kinematic engines have been demonstrated in the power range 0.1–500 kW. In free piston engines the reciprocating movement of the power piston is used to drive a linear electric generator. These engines are comparatively simple in mechanical terms and require little maintenance. Free piston engines have been demonstrated in the power range 0.05–3 kW [8].

The main feature that distinguishes modern Stirling engines from competing technologies such as internal combustion engines and fuel cells is the flexibility of the heat source. Stirling engines can be heated by concentrated sunlight, waste heat, and, depending on an appropriate burner design, a host of different fuel types. Stirling engines are therefore well placed for applications involving:

- concentrated sunlight: solar dish Stirling engines;
- combustion of biomass: biogas or oil fired Stirling engines.

Because Stirling engines are comparatively quiet and the external burners can operate with very low emission levels Stirling engines are also suitable for use in the built environment. As a result there has been an extensive effort in the past decade to develop Stirling engines custom made for micro-cogeneration systems. The Stirling engine technology is still not quite developed and represent an emerging technology although some applications already exist. They are commercially available as 55-kW units and are projected to be available in 150- to 300-kW sizes. They are also manufactured for much smaller powers of 1 kWe (or less) to a few kWe. The units generally available for micro-CHP are 1–9 kWe, with overall efficiencies in excess of 90%. Emissions from current Stirling burners can be ten times lower than that emitted from gas Otto engines with catalytic converter, making the emissions generated from Stirling engines comparable with those from modern gas burner technology. Stirling engines, being external combustion machines, have a number of advantages in terms of reliability and performance and ultimately should have a cost between that of spark and compression ignition automotive units, although current capital costs are high. Service intervals of be-

Table 8.4 Stirling engine manufacturers [8]

Manufacturer/location	
DTE Energy Technologies	55 kW–1 MW, overall efficiency of 84%
Kockums AB Sweden	
Sigma Elektroteknisk AS, Norway	3 kW$_e$ electrical output, 9 kW thermal output, Electrical efficiency > 25%
SOLO Kleinmoteren GmbH, Germany	2–9 kW electrical output, 8–24 kW thermal output, overall efficiency 92–96%
Stirling Energy Systems, Phoenix, AZ, USA	
Stirling Technology Co., Kennewick, WA, USA	
Stirling Technology, Inc., Athens, OH, USA	
Sunpower, Athens, OH, USA	7 kW electrical output
Tamin Enterprises, Half Moon Bay, CA, USA	
Whisper Tech Limited, New Zealand	Up to 1 kW electrical output, 7.5–12 kW thermal output

tween 3,500 h and 5,000 h (equivalent to more than 1 year's economic operation) are expected compared with 750–1000 h for IC engines. (Claims of longer service intervals are normally based upon oversized components such as large oil reservoirs which cannot be applied in normal domestic systems.) Life expectancy should be 50–60,000 h compared with 10,000 h for an IC engine. Table 8.4 presents some of the Stirling engine manufacturers (some in the development stage).

8.2.3 Fuel Cells for Micro-CHP

A fuel cell is an electrochemical device for the direct conversion of the chemical energy of hydrogen into electricity, heat and water vapor. This conversion can be done with very high electrical efficiency (35–55%) and with minimum environmental intrusion. These two aspects have rendered fuel cells the most likely energy conversion devices in the medium to long term, in both transport and stationary applications and for all power ranges. The capability of fuel cells to operate with a variety of fuels is of particular interest to the present study, since this makes them a favorite candidate for the optimal use of biofuels.

The basic operation of a fuel cell is exactly the opposite of electrolysis. In an electrolyzer, an electric current is passed through water which is broken into oxygen and hydrogen. In a fuel cell hydrogen and oxygen are combined producing an electric current and water. The principle was demonstrated by Sir W. Grove in 1839. Fuel cells operate very much like batteries, the main difference being that in batteries the reactants are stored within the battery itself and are limited by its size, while in a fuel cell they are stored externally and energy can be produced as long as fuel is fed to the anode and an oxidant to the cathode. The most common types of fuel cells are:

- Proton exchange membrane fuel cells (PEM) where the electrolyte is an ion exchange membrane. PEM fuel cells have the potential for the low costs that would make them suitable for home, farm, and similar small applications. A particular example is the direct methanol fuel cell (DMFC).
- Alkaline fuel cells (AFC), where the electrolyte is an 80% concentrated solution of KOH. They were the first in the market in spacecraft applications and have thermal efficiencies of up to 70%, but are typically too expensive for commercial use.
- Phosphoric acid fuel cells (PAFC) where the electrolyte is 100% concentrated phosphoric acid. They are already commercially viable for some applications and can approach thermal efficiencies of 85% if the steam byproduct is applied rather than wasted.
- Molten carbonate fuel cells (MCFC) where the electrolyte is alkali carbonates that in the high operating temperatures of this fuel cell (1,200°F) form molten salts. They are probably limited to industrial applications.
- Solid oxide fuel cells (SOFC) where the electrolyte is a solid, non-porous metal oxide. SOFC operate around 1,800°F and are thus probably limited to industrial applications and large power plants.

Fuel cells are expensive to build since demand has not reached the level that would allow mass production, meaning many devices are still built by hand. Additionally, some fuel cells use expensive materials, such as platinum. The average cost of fuel cells (depending on type and technology) for micro-CHPapplications is 5,000–15,000 €/kW.

Even though reliability is potentially higher than that of competing technologies, currently the reliability of fuel cells and their operating life is lower, due to the immaturity of the technology. Guarantees offered by most fuel cell manufacturers are limited to 1 year or 1,500 operating hours; therefore long term O&M costs are a significant cost factor [8].

The cost of integration of fuel cells into an existing micro-CHP energy system is considerable compared to other conventional solutions. In most cases the fuel that will drive fuel cells (pure hydrogen, biogas, other hydrocarbons) is not available on site and should also be purified so as not to poison the fuel cell catalyst. Therefore a drying and purification unit should be integrated into the energy system and this naturally increases the overall cost. In the case where fuel cells are driven by pure hydrogen, an electrolyzer with a considerable cost should also be added. These costs are eliminated when fuel cells driven by available on-site natural gas are used. Moreover, most fuel cells with a small to medium capacity, suitable for integration in micro-CHP energy systems deliver DC current, and therefore a DC/AC inverter and other power electronics with a considerable capital cost should also be added. In addition, safety precautions that should be taken into consideration in the presence of hydrogen also increase the cost of integration of fuel cells into an existing micro-CHP energy system.

Fuel cell systems do not involve the combustion processes associated with reciprocating internal combustion engines and micro-turbines. Consequently, they

Table 8.5 Estimated fuel cell emission characteristics [3, 9]

Fuel cell type	PEMFC	PEMFC	PAFC	SOFC	MCFC
Nominal electrical capacity (kW_{el})	10	200	200	100	250
Electrical efficiency (%) HHV	30	35	36	45	46
Emissions					
NO_x (ppmv at 15% O_2)	1.8	1.8	1.0	2.0	2.0
CO (ppmv at 15% O_2)	2.8	2.8	2.0	2.0	2.0
Unburnt hydrocarbons (ppmv at 15% O_2)	0.4	0.4	0.7	1.0	0.5
CO_2 (lb/MW h)	1360	1170	1135	910	950
Carbon (lb/MW h)	370	315	310	245	260

have the potential to produce fewer emissions. The major source of emissions is the fuel processing subsystem because the heat required for the reforming process is derived from the anode-off gas that consists of about 8–15% hydrogen, combusted in a catalytic or surface burner element. The temperature of this lean combustion process, if maintained below 1000°C, prevents the formation of oxides of nitrogen (NO_x). In addition, the temperature is sufficiently high for the oxidation of carbon monoxide (CO) and unburnt hydrocarbons. An absorbed bed helps in removing other pollutants such as oxides of sulfur (SO_x) (see Table 8.5).

8.2.4 Biofuels for Micro-CHP

Biomass can be used as a fuel for micro-CHP installations after pre-treatment and conversion into an applicable fuel. The conversion of the raw biomass into an applicable fuel can take place coupled with the final application at one location, or uncoupled with conversion and application at different locations. Uncoupled conversion has the advantage that large scale biomass pre-treatment and conversion technologies can be used; a disadvantage is the need for a distribution system for the final product, although it must be recognized that in coupled systems biomass transportation to the site also often takes place. Table 8.6 provides an overview of possible biomass-to-application chains for coupled supply chains that could be suitable for the current study [10].

Table 8.6 Coupled biomass-to-application chains suitable for micro-CHP

Biomass product category	Conversion technology	Product	Application
Solid biomass	Combustion	Hot flue gas	Stirling engine
	Gasification	Producer gas	Micro-turbine, Fuel cell
Liquid biomass	–	–	Micro-turbine
Gaseous biomass	Anaerobic digestion (manure)	Biogas	Micro-turbine

Examples of coupled biomass-to-application chains are the conversion of solid biomass, *i.e.*, wood chips, pellets, or briquettes into heat and application in a Stirling engine. For use in a Stirling engine a clean gaseous heat flow of a temperature of more than 900°C is needed. Anaerobic digestion of manure is a common CHP application, usually in the range of 35–300 kWe, so it can be regarded either as micro-CHP or small scale CHP. Anaerobic digestion of sludge and landfill gas extraction and electricity are usually performed at capacity levels substantially higher than 50 kWe. Liquid biomass like biodiesel, bio-ethanol, pure vegetable oil, and pyrolysis oil are all the result of conversion and upgrading of the biomass at a central facility site.

8.3 Market Survey on Future Penetration of Biofuel Micro-CHP in Europe

A market survey on the possible future penetration of the three biofuel micro-CHP technologies under examination, namely micro-turbines, Stirling engines, and fuel cells has been carried out for 25 European countries (EU-25): Belgium, Czech Republic, Denmark, Germany, Estonia, Greece, Spain, France, Ireland, Italy, Cyprus, Latvia, Lithuania, Luxembourg, Hungary, Malta, Netherlands, Austria, Poland, Portugal, Slovenia, Slovakia, Finland, Sweden, and United Kingdom. Table 8.7 summarizes the basic technological and economic parameters that have been used for the assessment of the market potential of the three technologies. The unit sizes, electric, and thermal efficiencies are based on the available data on the state of the art micro-CHP technologies that have been presented in the previous sections, while unit prices include the overall cost for the micro-CHP system together with the costs pertaining to the fuel system, *e.g.*, fuel drying and purification unit for a fuel cell, *etc.* The assumed prices, especially for the Stirling engine and fuel cell units, also reflect the fact that the engines will become cheaper as the market size increases, but they are both considerably higher than the micro-turbine unit price.

Due to lack of detailed data concerning the actual number of boilers in the countries under survey that could be replaced by micro-CHP units, the study is based upon the calculation of a number of "typical" boilers in each country and the corresponding full-load hours of operation per year. In order to evaluate these figures, the study has been based on available data from Eurostat concerning the

Table 8.7 Basic technological and economic assumptions for the three technologies

Unit type	Unit size (kW_{el})	Electric efficiency (%)	Thermal efficiency (%)	Unit price (Euro/kW)	Maintenance costs (Euros/kWh)
Micro-turbine	18.0	25.0	60.0	1,000	0.01
Stirling engine	2.0	20.0	60.0	1,800	0.02
Fuel cell	6.0	40.0	50.0	3,000	0.02

yearly fuel consumption and electricity consumption for households, coupled with available data on total population per country, average number of persons per household, and building distributions. The Eurostat final energy consumption data from households that have been used for the calculations are presented in Table 8.7. Based on the population of each country, the household fuel energy consumption and electricity consumption per capita is calculated. Eurostat provides the household energy consumption per capita, which also includes electricity consumption for heating and other appliances. By subtracting the fuel energy consumption from the Eurostat energy consumption, the electricity consumption per capita for space heating/water heating is obtained. Next, the total efficiency for household fuel consumption is calculated per country for the fuel mixture used (electricity for space/water heating included), by assuming a fuel efficiency of 70% for gas + petroleum, 40% for solid fuels, 20% for biomass, and 95% for electricity. The above-mentioned calculations are presented in Tables 8.8 and 8.9.

Available data on the average number of persons per household and percentage distribution of buildings per category (detached and semi-detached single family house, building with less than 10 dwellings, building with more than 10 dwellings), have been used to calculate the number of buildings per category and persons per building, with the assumption of a fixed number of families per type of building (1, 6, 18 for each type of building respectively, uniform throughout Europe since no other data are available). Data on average persons per family and distribution of buildings were available only for the EU-15. For the rest of the EU-25, persons per family have been assumed higher, due to a lower GDP, and the buildings distribution was chosen the same as in neighboring EU-15 countries. The aforementioned data are summarized in Table 8.10

The above calculated data on the energy consumption and efficiency, buildings and persons per building category (Table 8.11) are used in order to calculate a number of "typical" size boilers per building type and fuel category, the average full load hours a micro-CHP unit of a given output can be operated, based on the estimated electricity demand of the building, and the heat requirements per building. The number of boilers of a given fuel type X is given by the following formula:

$$Boilers\ of\ a\ fuel\ type\ X = \frac{efficiency\ of\ fuel\ X\ input\ fuel\ X}{average\ efficiency\ X\ total\ input\ fuel} \cdot total\ number\ of\ households$$

The efficiencies of fuel X used in the calculations are 70% for gas + petroleum, 40% for solid fuels, and 20% for biomass. The calculated numbers of boilers for the 25 countries under investigation for each building category and type of fuel is presented in Table 8.11.

The estimation of the market potential for the three types of micro-CHP units is conducted next. The units were chosen so that each unit size can cover the needs of a particular type of building: the 2-kW Stirling engine for detached and semi-detached single-family houses, the 6-kW fuel cell for buildings with less than ten dwellings and the 18-kW micro-turbine for buildings with more than ten dwellings and therefore each technology is examined in terms of possible market penetration in the particular market segment. The estimation has been based on the combina-

tion of two factors: the first one is a "marketability factor", defined by the ratio
capital cost/GDP per building, the latter obtained from Eurostat. This factor has
been used in order to represent the maximum capital investment relative to the
income in a particular country that the residents of a building would be willing to
pay for the purchase of micro-CHP equipment. The second is the payback factor,
representing the economic usefulness of the capital investment. For each of the
two factors, a maximum acceptable value is set rendering the factor 0, while for
values less than the maximum, there is a linear variation of the factor from 0 to 1.
For the present study, the maximum values of the factors have been set to 10% of
the GDP (per building) and 7 years for the payback period.

The payback period is calculated from the capital costs and the annual profit
from the investment in each case. The annual profit equals to the annual costs for
operation of the units (CHP fuel + maintenance) minus the cost of displaced fuel
and cost of electricity displaced. The cost of displaced fuel is calculated by the
following equation:

Heat output (CHP) × Price of fuel displaced/Thermal efficiency of fuel displaced,

where

Heat output (CHP) = Full load hours × Unit size × Thermal efficiency/Elec.
efficiency

The calculation of the annual profit includes an adjustment if the heat coverage
of the micro CHP system is greater or equal to 100%, to account for the actual
heat requirement of the building. If it is greater than unity, only 100% of the heat
requirement is included in the profit calculation, assuming that the rest is not used.
For the above calculations, the operating and economic data of the units (Table 8.7), the estimated full load hours of operation of the units, and the prices of
electricity and fuels (cogeneration fuel and fuel displaced) have been used.

As concerns the electricity price and fuel prices, data were acquired from the
Eurostat database for the various European countries. The electricity prices were
found in the form of yearly prices per kWh including taxes for different ranges of
electricity consumption. The biomass price has been assumed equal to
0.01 Euros/kWh$_{th}$. Table 8.12 presents basic input data for the above describe
calculations, while Table 8.13 presents the results of the survey for the possible
future penetration of the selected biofuel micro-CHP systems.

The results show a potential of ca. 6 million units of 2-kW Stirling engines for
detached and semi-detached single-family houses, 180,000 units for 6-kW fuel
cells for buildings with less than 10 dwellings, and 1.2 million units of 18-kW
micro-turbine units for buildings with more than 10 dwellings for the EU-25, with
an annual electricity fed to the grid of ca. 122 TWh. The results also include an
estimation of the required annual biomass quantities for operation of the installed
units, for typical biomass, for each country under investigation. The availability of
the quantities is an additional another factor that could affect the market potential.
Nevertheless, the current survey shows that there is indeed a significant potential
in Europe for application of decentralized micro-CHP technologies in the future.

Table 8.8 Final energy consumption from households for various fuels and electricity (source: Eurostat)

Country	Solid fuel (KTOE – kilo Tons of Oil Equivalent)	Hard coal (ktons)	Coke (ktons)	Patent fuel (ktons)	Lignite and peat (ktons)	Brown coal briquettes (ktons)	Petroleum products (KTOE)	Natural Gas (KTOE)	Renewables (wood, wood waste, etc.) (KTOE)	District heating (TJ)	Electricity (GWh)
Belgium	225	304	7	7	0	5	4,304	3,457	152	15	25,921
Czech Republic	861	1,816	157	0	6	123	6	2,230	52	1,282	14,121
Denmark	20	33	0	0	0	1	1,338	659	551	1,502	10,181
Germany	1,203	440	212	105	25	1,484	27,408	24,725	4,048	0	131,172
Estonia	35	39	0	0	4	18	129	37	328	393	1,584
Greece	25	2	0	0	0	64	3,756	9	704	28	15,775
Spain	67	120	0	0	0	0	5,039	2,513	2,063	0	50,636
France	545	668	39	139	1	0	14,249	8,407	7,278	0	132,998
Ireland	447	288	0	46	622	261	1,068	475	47	0	7,449
Italy	16	1	23	0	0	0	6,770	14,167	1,426	0	62,957
Cyprus	0	0	0	0	0	0	85	0	45	0	1,157
Latvia	55	81	0	0	0	0	91	79	848	465	1,317
Lithuania	81	120	0	0	14	11	38	112	509	552	1,811
Luxembourg	4	0	0	0	0	8	295	228	15	16	737
Hungary	276	387	9	0	215	61	293	3,282	345	607	10,440
Malta	0	0	0	0	0	0	18	0	0	0	500
Netherlands	33	32	0	0	23	0	646	7,826	155	158	22,800
Austria	223	92	178	1	45	65	2,179	1,458	2,070	508	15,685
Poland	6,330	10,184	685	0	330	1	3,935	3,036	3,104	4,968	21,659
Portugal	0	0	0	0	0	0	789	147	1,150	7	11,382
Slovenia	2	3	0	0	2	0	654	65	361	97	2,704
Slovakia	397	88	0	0	11,02	1	84	1,603	3	782	4,907
Finland	25	4	0	0	94	0	1,732	26	1,182	1,433	19,940
Sweden	0	0	0	0	0	0	1,350	44	877	2,404	41,416
UK	1,511	1,868	178	391	0	0	1,614	29,128	357	12	11,4534

Table 8.9 Calculations of household energy consumption and efficiency

Country	Population (thousands)	Calculated household fuel energy consumption per capita (kgOE)	Household energy consumption per capita (kgOE) – from Eurostat tables	Calculated household electricity consumption per capita (kWh)	Calculated household electricity consumption for heating per capita (kWh)	Total input fuel (KTOE)	Calculated total efficiency for household fuel consumption (%)
Belgium	10,333	788	898	2,509	1,284	9,279	71.5
Czech Republic	10,205	312	524	1384	2,471	5,347	74.9
Denmark	5,411	481	785	1,882	3,534	4,248	73.2
Germany	82,489	696	732	1,590	423	60,382	67.3
Estonia	1,359	396	702	1,166	3,558	954	62.8
Greece	11,003	408	445	1,434	425	4,896	64.7
Spain	41,201	235	311	1,229	884	12,814	67.9
France	59,486	512	632	2,236	1,391	37,595	64.6
Ireland	3,932	518	667	1,894	1,733	2,623	69.6
Italy	57,157	392	478	1,101	1,006	27,321	71.9
Cyprus	710	183	328	1,630	1,685	233	71.4
Latvia	2,339	463	611	563	1,717	1,429	45.4
Lithuania	3,436	219	400	527	2,104	1,374	61.2
Luxembourg	446	1,216	1,383	1,652	1,942	617	71.6
Hungary	10,159	414	548	1,028	1,553	5,567	71.6
Malta	396	45	156	1,263	1,286	62	87.7
Netherlands	16,149	537	628	1,412	1,064	10,142	72.8
Austria	8,053	738	875	1,948	1,594	7,046	58.3
Poland	38,174	433	471	567	444	17,980	53.0
Portugal	10,368	201	301	1,098	1,161	3,121	59.9
Slovenia	1,995	543	594	1,355	588	1,185	56.9
Slovakia	5,379	392	549	912	1,831	2,953	73.2
Finland	5,201	577	997	3,834	4,888	5,185	69.1
Sweden	8,925	261	854	4,640	6,898	7,622	81.8
UK	59,234	551	744	1,934	2,250	44,070	75.1

Table 8.10 Data and calculations of buildings/persons

Country	Average persons per household	Detached and semi-detached single family house (%)	Building with less than 10 dwellings (%)	Building with more than 10 dwellings (%)	Detached and semi-detached single family houses (Buildings)	Building with less than 10 dwellings (Buildings)	Building with more than 10 dwellings (Buildings)	Persons per Detached and semi-detached single family house	Persons per Building with less than 10 dwellings	Persons per Building with more than 10 dwellings
Belgium	2.4	77.9	11.3	10.8	3,355,692	80,902	25,795	2.40	14.40	43.20
Czech Republic	3.0	55.4	2.9	41.7	1,884,257	16,675	78,742	3.00	18.00	54.00
Denmark	2.2	57.4	14.2	28.4	1,430,123	59,079	39,386	2.17	13.02	39.06
Germany	2.2	39.2	38.2	22.5	14,703,922	2,389,387	469,709	2.20	13.20	39.60
Estonia	3.0	55.4	2.9	41.7	25.0926	2,221	10,486	3.00	18.00	54.00
Greece	2.6	45.6	32.8	21.6	1,929,259	231,649	50,709	2.60	15.60	46.80
Spain	3.0	36.8	17.2	46.1	5,049,142	392,711	351,570	3.00	18.00	54.00
France	2.4	57.4	14.2	28.4	14,215,404	587,246	391,497	2.40	14.40	43.20
Ireland	3.0	93.1	3.9	2.9	1,220,719	8,566	2,142	3.00	18.00	54.00
Italy	2.6	31.9	35.8	32.4	7,004,534	1,311,105	395,128	2.60	15.60	46.80
Cyprus	3.0	45.6	32.8	21.6	107,892	12,955	2,836	3.00	18.00	54.00
Latvia	3.0	55.4	2.9	41.7	431,874	3,822	18,048	3.00	18.00	54.00
Lithuania	2.7	33.1	4.8	62.1	421,228	10,181	43,904	2.70	16.20	48.60
Luxembourg	2.5	66.7	20.1	13.2	118,933	5,976	1,312	2.50	15.00	45.00
Hungary	3.0	45.6	32.8	21.6	1,543,770	185,363	40,577	3.00	18.00	54.00
Malta	3.0	45.6	32.8	21.6	60,176	7,225	1,582	3.00	18.00	54.00
Netherlands	2.3	67.2	5.4	27.5	4,715,288	63,100	107,079	2.30	13.80	41.40
Austria	2.4	43.6	15.2	41.2	1,463,883	84,982	76,758	2.40	14.40	43.20
Poland	3.1	55.4	2.9	41.7	6,887,753	60,954	287,836	3.07	18.42	55.26
Portugal	2.9	68.1	20.1	11.8	2,436,024	119,757	23,367	2.90	17.40	52.20
Slovenia	3.0	43.6	15.2	41.2	290,123	16,842	15,212	3.00	18.00	54.00
Slovakia	3.0	43.6	15.2	41.2	782,240	45,411	41,016	3.00	18.00	54.00
Finland	2.1	55.4	2.9	41.7	1,371,879	12,141	57,330	2.10	12.60	37.80
Sweden	1.9	55.4	2.9	41.7	2,601,974	23,026	108,735	1.90	11.40	34.20
UK	2.4	81.9	11.3	6.9	20,204,408	463,774	94,099	2.40	14.40	43.20

Table 8.11 Calculated number of "typical size" boilers per building category

Country	Boilers with solid fuels			Boilers with petroleum products and natural gas			Boilers with renewables (wood, wood waste, *etc.*)		
	Detached and semi-detached single-family house	Building with less than ten dwellings	Building with more than ten dwellings	Detached and semi-detached single-family house	Building with less than ten dwellings	Building with more than ten dwellings	Detached and semi-detached single-family house	Building with less than ten dwellings	Building with more than ten dwellings
Belgium	45,504	1,097	350	2,746,809	66,223	21,115	15,370	371	1,182
Czech Republic	161,952	1,433	6,768	736,020	6,513	30,758	4,891	43	2,044
Denmark	3,679	152	101	642,682	26,550	17,700	50,674	2,093	13,956
Germany	174,137	28,297	5,563	13,206,151	2,146,000	421,863	292,979	47,609	93,590
Estonia	5,863	52	245	48,623	430	2,032	27,474	243	11,481
Greece	6,089	731	160	1,604,485	192,653	42,173	85,729	10,294	22,533
Spain	15,552	1,210	1,083	3,067,627	238,593	213,598	239,435	18,623	166,718
France	127,566	5,270	3,513	9,280,435	383,380	255,587	851,765	35,187	234,579
Ireland	119,614	839	210	722,403	5,069	1,267	6,288	44	110
Italy	2,282	427	129	5,226,172	978,232	294,810	101,702	19,037	57,371
Cyprus	0	0	0	38,617	4,637	1,015	5,841	701	1,535
Latvia	14,652	130	612	79,101	700	3,306	112,955	1,000	47,204
Lithuania	16,222	392	1,691	52,409	1,267	5,463	50,968	1,232	53,124
Luxembourg	431	22	5	98,550	4,952	1,087	808	41	89
Hungary	42,783	5,137	1,125	969,888	116,456	25,493	26,739	3,211	7,028
Malta	0	0	0	13,993	1,680	368	0	0	0
Netherlands	8,432	113	191	3,788,244	50,694	86,027	19,802	265	4,497
Austria	31,779	1,845	1,666	907,136	52,661	47,565	147,495	8,562	77,338
Poland	1,831,381	16,207	76,533	3,529,499	31,235	147,496	449,021	3,974	187,644
Portugal	0	0	0	853,865	41,977	8,191	299,885	14,743	28,766
Slovenia	344	20	18	216,557	12,572	11,355	31,073	1,804	16,293
Slovakia	57,456	3,335	3,013	427,381	24,811	22,409	217	13	114
Finland	3,827	34	160	470,971	4,168	19,682	90,472	801	37,808
Sweden	0	0	0	407,316	3,605	17,022	73,235	648	30,604
UK	369,126	8,473	1,719	13,142,433	301,673	61,209	43,606	1,001	2,031

Table 8.12 Basic economic input for the calculations

Country	GDP (MEuros)	GDP per capita (Euros)	Electricity price (Euro/kWh) (Eurostat)	N.G. fuel price (Euros/kWh) (Eurostat)
Belgium	245,343	23,744	0.0926	0.0319
Czech Republic	44,450	4,356	0.0491	0.0227
Denmark	161,384	29,825	0.2075	0.0453
Germany	2,076,860	25,177	0.1008	0.0366
Estonia	4,000	2,943	0.0615	0.0141
Greece	115,046	10,456	0.0579	0.0450
Spain	566,378	13,747	0.0707	0.0369
France	1,400,756	23,548	0.0930	0.0324
Ireland	92,225	23,455	0.0777	0.0317
Italy	942,346	16,487	0.1834	0.0450
Cyprus	8,656	12,192	0.0708	0.0450
Latvia	5,166	2,209	0.0463	0.0139
Lithuania	18,084	5,263	0.0647	0.0165
Luxembourg	20,051	44,957	0.0852	0.0276
Hungary	44,641	4,394	0.0809	0.0194
Malta	3,117	7,871	0.0958	0.0450
Netherlands	386,785	23,951	0.1222	0.0347
Austria	209,993	26,076	0.1125	0.0321
Poland	222,486	5,828	0.0535	0.0223
Portugal	101,933	9,832	0.0874	0.0423
Slovenia	20,062	10,056	0.0723	0.0282
Slovakia	19,256	3,580	0.0772	0.0246
Finland	129,171	24,836	0.0683	0.0266
Sweden	229,349	25,697	0.1170	0.0422
UK	1,051,071	17,744	0.0647	0.0233

Table 8.13 Basic results on market penetration of selected biofuel micro-CHP technologies

Country	2-kW Stirling		6-kW fuel cell		18-kW micro-turbine		Annual biomass consumption (ktonnes) LHV = 8 MJ/kgr – 30% moisture
	No. of Units	Electricity generation (GWh)	No. of Units	Electricity generation (GWh)	No. of Units	Electricity generation (GWh)	
Belgium	602,506	3,627	10,326	373	15,426	1,672	11,590
Czech Republic	0	0	0	0	5,171	386	696
Denmark	192,791	787	8,752	214	19,912	1,463	4,646
Germany	1,915,663	6,702	3,629	76	302,099	19,023	49,407
Estonia	0	0	0	0	0	0	0
Greece	0	0	0	0	20,447	1,372	2,470
Spain	34,821	128	58	1	151,583	10,060	18,398
France	1,455,850	7,812	49,880	1,606	268,662	25,949	66,092
Ireland	216,919	1,233	663	23	1,099	112	3,002
Italy	351,611	1,007	94,269	1,620	200,758	10,349	22,716
Cyprus	364	2	146	4	833	73	141
Latvia	0	0	0	0	0	0	0
Lithuania	0	0	0	0	58	1	3
Luxembourg	19,369	80	3	0	781	58	285
Hungary	0	0	0	0	3,410	189	341
Malta	0	0	0	0	172	12	21
Netherlands	513,722	1,668	1,214	24	55,920	3,269	9,663
Austria	196,473	918	9,021	253	67,638	5,691	12,595
Poland	0	0	0	0	13,579	426	766
Portugal	0	0	0	0	4,010	230	414
Slovenia	0	0	0	0	5,359	392	706
Slovakia	0	0	0	0	967	48	86
Finland	94,196	758	575	28	27,444	3,977	8,897
Sweden	94,633	834	1,091	58	23,277	3,694	8,592
UK	272,304	1,264	1	0	33,084	2,764	7,818
Total	5,961,222	26,821	179,628	4,280	1,221,689	91,211	229,342

8.4 Conclusions

The current work has presented results from the European co-ordination action project MICROCHEAP that was intended to bring together industrial specialists and research experts to focus entirely on renewable micro-CHP technology, co-ordinate and steer research in this field, and highlight the most promising technologies with the highest potential for market penetration in existing and future market conditions. The state of the art in technological options in the field of renewable micro-CHP with biofuels with regards to technology, cost, and environmental impacts has been presented in detail and results of a market survey have been provided, concerning the possibility of future penetration of three selected biofuel micro-CHP technological options in Europe. The results have demonstrated a significant potential for future application of the technology in households. Further technological developments and capital investment reductions might make such installations even more feasible in the near future.

References

1. DTI (2006) Microgeneration strategy
2. Bruno JC, Valero A, Coronas A (2005) Performance analysis of combined microgas turbines and gas fired water/LiBr absorption chillers with post-combustion. Applied Thermal Engineering 25:87–99
3. Onovwiona HI, Ugursal VI (2006) Residential cogeneration systems: review of the current technology. Renew Sustain Energy Rev 10:389–431
4. Environmental Protection Agency, Climate Protection Partnership Division (2002) Technology characterization: microturbines
5. Pilavachi PA (2002) Mini and micro-gas turbines for combined heat and power. Int J Appl Thermal Eng 22:2003–2014
6. Office of Industrial Technologies, Office of Energy Efficiency and Renewable energy, U.S. Department of Energy (1999) Review of combined heat and power technologies
7. Joerss W, Joergensen BH, Loeffler P *et al.* (2002) Biogas applications for large dairy operations: decentralised generation technologies potentials, success factors and impacts in the liberalised EU energy markets. CE Final Report NNE5-1999-593
8. MICROCHEAP (2005) Report WP5 – Investigation of links between renewable energy systems and micro-CHP
9. Nexus Group (2002) Technology characterization—micro-turbines. Environmental Protection Agency, USA
10. MICROCHEAP (2005) Report WP2 – D3: Literature Search

Chapter 9
Ash Formation, Slagging and Fouling in Biomass Co-firing in Pulverised-fuel Boilers

M.K. Cieplik, L.E. Fryda, W.L. van de Kamp and J.H.A. Kiel

Abstract This chapter gives an overview of the main ash formation and deposition mechanisms for various relevant biomass fuels, also in blends with selected coals, in pulverised-fuel (PF) boilers. The chapter is divided into three sections. In the first, a general outline of the ash formation mechanisms is given. The second section gives a review of experimental and analytical techniques for the (lab-scale) characterisation of fuels, with the emphasis on the ash-forming elements contents and fate during the combustion. Fuel reactivity and burnout, devolatilisation behaviour, N-release and slagging and fouling propensity are discussed. Also, a detailed overview of the experimental conditions and their relevance for the existing as well as the future technologies is given. Further an outline of diagnostic techniques for the in-boiler characterisation of slagging and fouling is issued. In the third section, key ash-formation phenomena are discussed for various pure biomass fuels and selected typical coals. This is done on the basis of exemplified results, generated with the techniques discussed in the foregoing section. For slagging and fouling this is also backed up with data from full-scale diagnostic measurements.

9.1 Ash Formation Mechanisms – Outline

Coal, biomass and wastes are extensively used in a multitude of technologies to generate heat and power. In the most basic terms, to produce energy means to release the chemical energy from the organic, convertible matter in the said fuels,

M.K. Cieplik (✉)
Energy research Centre of the Netherlands (ECN), Biomass, Coal and Environmental Research,
P.O. Box 1, 1755 ZG Petten
Tel: +31 224 56464700
e-mail: Cieplik@ecn.nl

consisting in the major part of hydrocarbons and carbohydrates. However this convertible matter is accompanied by mineral matter. Although this *ash* is present in much lower concentration than the carbonaceous matrix, it forms a focal point in the design and operation of power plants. Even though often thermally relatively stable, the ash undergoes in the flame/furnace a multitude of physical processes, as it devolatilises, fragments and condenses partly in due course of thermal conversion of the fuel. These physical transformations during combustion of solid fuel are schematically shown in Figure 9.1 [1–3].

This inorganic residue after combustion for a major part travels as a suspension of fine particulate towards the stack, along with the flue gas and it can potentially create various problems such as near-burner slagging, boiler fouling, corrosion and erosion [4]. It can also affect emissions in various ways.

The two important physical transformations are fragmentation and vaporisation. The vaporised minerals chemically react with other volatiles, melts or solids during combustion. The physical and chemical transformations during thermal conversion of solid fuel are time-dependent and very difficult to understand, as a continuous process. They depend on several fuel characteristics, *e.g.* fuel mineral matter composition, ash levels in fuel, fixed carbon, volatile matter, mineralogy (particularly the levels of included or excluded mineral phases), char reactivity and char morphology, density and particle size. Also the operating conditions are of major importance for the said transformations. Crucial parameters include the type of combustion system (air staging), temperature, pressure, heating rate and residence time, as they affect chemical equilibria of numerous gaseous species as well as reactivities of gaseous, liquid and solid slag phase-bound minerals.

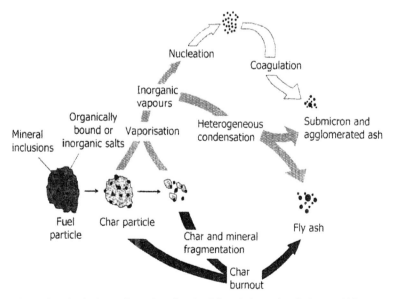

Figure 9.1 Physical transformations involved for ash formation during coal/biomass combustion

9.2 Parameters of Importance for Mineral Transformations

9.2.1 Fuel Mineral Matter Composition and their Association

Coal, biomass or their blends can lead to different ash formation mechanisms during pulverised-fuel (PF) combustion, as the fuel mineral matter composition and association vary greatly in different fuels. The mineral matter in the fuel may be present as free ions, salts, organically bound or hard (included or excluded) minerals. The younger fuels contain a major fraction of volatile compounds (and less minerals) compared to older fuels. Alkalis in the younger fuels such as woody biomass, but also low rank coals, mainly remain in included minerals as free ions, salts and organically-associated inorganics. Readily available, these start vaporising at lower temperatures even before char burnout. It is highly probable that these alkaline metals will chemically react and in the later stage of the combustion process condense to nucleate and coagulate on each other or onto the wall of the furnace, to produce submicron ash. Other inorganics such as calcium and magnesium volatilise, fragment and/or coalesce [1, 5] only in part.

However, the vaporisation of the volatile inorganics is not limited to the split-second explosive heat up in the flame. During diffusion-limited char combustion, the interior of the particle becomes hot and fuel rich. The non-volatile oxides (*e.g.* Al_2O_3, SiO_2, MgO, CaO, Fe_2O_3) can be reduced to more volatile suboxide and elements, and partly vaporised. These can then re-oxidise while passing through the boundary layer surrounding the char particle, becoming highly supersaturated and nucleate homogeneously [6, 7]. Also, fragmentation behaviour is highly dependent on the general fuel rank. Baxter [8] studied three different ranks of pulverised coal (high-volatile bituminous, sub-bituminous and lignite). He observed that for high-volatile coal more than 100 fly ash particles were formed from a single 80-µm (initial diameter) char particle, whereas only 10 fly ash particles were produced from a single 20-µm (initial diameter) char particle. However, regardless of its initial size, fragmentation of lignite particles was far less extensive, with fragments per original char particle being less than five. Nonetheless, whatever the fragmentation characteristics, it has been verified by, amongst others, Buhre *et al.* [9], that formation of submicron ash particles during coal and biomass combustion is mainly due to condensation of evaporated species.

Ash melting behaviour, which is of crucial importance particularly for slagging and fouling, is affected by the elemental composition of ash, primarily the levels of alkali metals, phosphorus, chlorine, silicon and calcium, as well as the chemical concentration of the compounds.

If molten phases are formed, the fate of the alkalis is highly dependent on their original speciation. Thy [10] found that if alkali metals occur as network-modifying and charge-balancing cations in highly depolymerised melts, such as wood/bark, they are easily evaporated during prolonged heating and subsequently deposited on cooler heat exchangers downstream. In contrast, if the melt is highly polymerised such as in rice straw, where alkali metals occur as network-modifying cations, they are strongly retained in the polymerised network.

9.2.2 Mineralogy

The mineral composition of fuels also plays a critical role in various physical and chemical transformations. Excluded minerals are present in biomass as a result of contamination with soil during harvest or handling while presence in coal is due to mining or handling. Included minerals are the inorganics required by plants for growth and they therefore still show up in coal, of biomass origin after the geological processes of peatification and coalification [11]. It is quite obvious that levels of excluded minerals in most of clean biomass will be considerably less than in the coal derived from mines.

The included minerals have a higher tendency to remain in the char during combustion. Wigley [12] showed that coal particles that contain included mineral matter will have a greater specific heat capacity than particles consisting of organic material alone, so particles with included mineral matter would be expected to heat up and combust more slowly. On the other hand, the included minerals are generally more volatile than the excluded ones.

In the later stage of thermal conversion, due to exothermic reactions occurring in the char during combustion, the included mineral matter can reach very high temperatures, even above those of the surrounding gas. As included minerals are situated close to each other, reactions between them can easily take place. As a result, the included minerals may either appear as molten particles on a reducing char surface or as a lattice network in the char particle itself. As the char burnout proceeds, the minerals may coalesce onto a single particle or fragment into several small particles, which is basically the result of the difference of thermal expansion coefficients between included minerals and their organic matrix.

Included minerals can also affect the conversion kinetics of char, as the minerals may fuse and coat the surface of burning char particles, reducing the rate of char combustion. On the other hand, included mineral matter, particularly if containing potassium, may catalytically promote char oxidation, which is often seen in biomass.

Generally, excluded minerals (especially for coal) will reach lower temperatures than included minerals, and they will not be influenced by locally reducing environment. The transformation occurring in excluded minerals and the behaviour with regard to deposition may therefore significantly differ from the included minerals. Excluded minerals can either be carried through the combustion system with their original structure intact or they can melt and fragment. Dacombe [13], Liu [14] and many others explained that excluded minerals always fragment randomly due to thermal stress. Ten Brink [15] and Li [16] observed that calcite and pyrite as excluded minerals fragment at high temperature and high heating rate conditions, while siderite and ankerite grains do not fragment under the same conditions.

9.2.3 Particle Shape, Size and Density

Experimental and theoretical investigations indicate that particle shape, size and density influence particle dynamics, including drying rate, heating rate and reaction rate [17]. It is generally observed that spherical particles devolatilise more quickly compared to other shape particles. Mathews [18] observed that mineral matter and maceral composition of the char will be different at different size which can affect the devolatilisation rate. No *et al.* [19] and Dacombe [13] observed that large particles form more fragments than small particles due to a larger internal temperature gradient. Wigley [12] observed that decrease in char particle size leads to more complete combustion. The ash transport behaviour is naturally affected by particle size after combustion to a large extent as well. Large ash particles tend to impact boiler heat transfer surfaces by inertia, whereas fine ash particles tend to reach wall surfaces by thermophoresis or Brownian motion. For example, a 60-μm ash particle was estimated to reach the deposit surface almost three times faster compared with a 30-μm particle primarily due to inertial effects [14].

9.3 Analytical and Experimental Techniques for the (Lab-scale) Characterisation of Fuels

There is a variety of analytical techniques for the characterisation of the chemical and physical composition of fuels. Although nowadays analytical methods are often very accurate, they generally are only indicative of the fuels dynamic parameters, *i.e.* the behaviour of the fuel under the real conditions of the conversion technology of choice. The latter characterisation is attempted by means of well-controlled experimental techniques, often in combination with advanced fuel characterisation, beyond the scope of the standard industrial fuel analyses.

In this section, a brief overview of the regular fuel analyses is given, followed by an outline of a few advanced analytical as well as experimental techniques and installations for more complete and adequate fuel characterisation.

9.3.1 Analytical Methods for Fuel Characterisation

9.3.1.1 Conventional Fuel Analyses

Proximate, ultimate and physical analyses. Typically, fuel analyses consist of several analytical steps. First, in order to quantify the general composition of the fuel, proximate and ultimate analyses are performed, including parameters such as water and ash contents and general combustible matter composition (C, H, N, O, S), its volatility and calorific value. Physical parameters, such as Particle Size Distribution (PSD) are also determined. These basic parameters allow for technical assessment of the fuel's suitability for application in power generation, yet they do not give insights into the fuel's true conversion kinetics, nor the environmentally relevant parameters such as the conversion of fuel nitrogen into NO_x.

Fuel ash analyses. In order to get insights into fuel ash behaviour in the furnace and its possible applications after thermal conversion (*e.g.* in the cement industry and or road construction), the ash needs to be analysed in more detail. The depth of fuel ash analyses depends on the sort of fuel, the specific conversion technology and the foreseen final application for the residual ash. Often the analyses begin with physical characterisation of the ash, for example by means of the standard Ash Fusion Test (AFT), which is meant to shed light on the melting properties of the ash. The elemental composition of the fuel ash is also often determined. Commonly analysed ash-forming elements in these fuels are silicon, aluminium, iron, calcium, magnesium, manganese, sodium, potassium, phosphorus, sulphur and chlorine. Some mandatory trace elements, of crucial importance for the toxicity of the emitted flue gases (Co, Cr, Hg, Mn, Mo, Ni, Pb, Sb, Se, Tl and V), are analysed as well. However, just as in the case of the proximate and ultimate analyses, the elemental composition analyses do not tell much about the fuel ash behaviour in the furnace in terms of elemental reactivity and partitioning, both of crucial importance for phenomena such as slagging and fouling.

9.3.1.2   Advanced Fuel Analyses and Predictive Models (Exemplified)

CCSEM Mineralogical analyses. As outlined in the foregoing sections, the mineralogical composition of fuels may be of crucial importance for their behaviour in a combustion environment. Thus, quantifying the distribution of elements in various mineralogical constituents of the fuel has a high added value as compared to the regular elemental analyses. One of the advanced techniques for the mineralogical analyses is Computer Controlled Scanning Electron Microscopy (CCSEM). The procedure [20] includes automated (particle-to-particle) analyses of the fuel embedded in a resin/wax blend, followed by computer-aided evaluation of the raw SEM/EDX data. Although not quite as accurate as ICP analyses, the output includes a particle-size-resolved elements classification into 25 + mineralogical phases. The technique can also be applied to differentiate between the excluded and the included mineral phases.

Chemical fractionation. CCSEM techniques are very well suited for fuels mineral phases characterisation. However, they are much less potent for the quantification of organically-bonded inorganic elements. Predicting the extent of the volatilisation of such inorganics into the flue gas can also be attempted by means of the chemical fractionation methods [4, 21], *e.g.* by treating the fuel with increasingly strong leaching chemicals. Ash forming elements, leachable in hydrochloric acid or not leachable at all, are likely to be constituents of relatively unreactive compounds. The elements leached out in the water and ammonium acetate solutions may, on the other hand, be considered as more reactive and may participate in reactions with bed ash particles and/or with fly ash particles in the flue gas channel or they may form submicron particles by homogeneous nucleation.

Thermodynamic and CFD modelling. Outputs of chemical fractionation analyses as well as a similar method [22] based on a pH static extraction can also be applied to unravel the physicochemical composition of the ashes, particularly the complexation and oxidation state of (trace) elements [23]. In order to do so, data from the exhaustive pH-static extraction are fed into computer models LeachXS and Orchestra, which translate the composition of the leachate back into the composition of solid phase. Thermodynamic modelling can, in some cases [24], be applied to model ash compositions and properties. Computer Fluid Dynamic (CFD) modelling is rapidly gaining ground nowadays as a way to predict and characterise various aspects of the fuel behaviour under certain combustion conditions or in a particular boiler design. Currently, dedicated sub-models are also being developed for the modelling of ash deposition phenomena [25]. However, the predictive value of such models is very dependent of the quality of the input parameters, also being the primary fuel characteristics.

9.3.2 Experimental Methods

As mentioned in the foregoing section, numerous analytical techniques in combination with advanced modelling can generate much insight into the composition of fuels and their ash-forming constituents. Nonetheless, as the analyses are static and thus not subjecting the fuels to the same dynamic combustion conditions that they would encounter in a real PF boiler, there has been numerous initiatives to try to mimic the combustion environment either on the lab-scale, or in down-scaled pilot installations. Also, sampling tools and techniques have been developed for in-boiler diagnostics, as described briefly in the next section.

9.4 Drop Tube Furnaces

The most prominent example of a lab-scale installation for dynamic fuel characterisation is a so-called Drop Tube Furnace (DTF). In its basic form, the installation constitutes a vertical reactor, which is externally heated, mostly by an electric furnace. Installations of such general layout have been utilised by many research institutes. A few, *e.g.* Casella/RWE N-Power (UK), INCAR (SP), CERCHAR (F) and ECN (NL), even attempted a harmonisation of this research technique [26]. Also nowadays new installations of this kind are being constructed [27]. The main advantage of the lab-scale DTF is its simplicity and flexibility in creating virtually any combustion environment and studying dynamically fuel conversion and ash formation. A classical DTF, however, also has many limitations. For example, the heating rates of a fuel particle falling freely in a heated tube are still at least one order of magnitude lower than in a real combustion system, where fuel particles are instantaneously subjected upon injection to the scourging heat of the PF flame. Nonetheless, with a few smart design features many of the said limitations can be overcome. A good example of such an advanced DTF is the Lab-scale Combustion Simulator, developed and optimised over the past decade by ECN.

9.4.1 Lab-scale Combustion Simulator (LCS)

A wide range of lab-scale co-firing tests have been performed in the ECN Lab-scale Combustion Simulator (LCS) [28], as depicted in Figure 9.2.

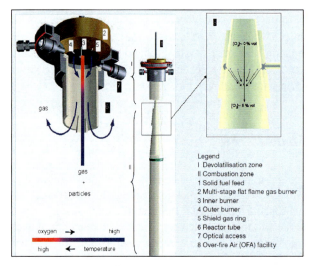

Figure 9.2 Schematic of the ECN's Lab-scale Combustion Simulator (LCS)

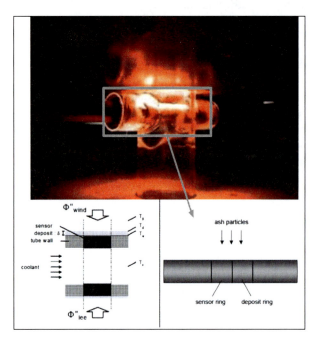

Figure 9.3 Schematic of the Horizontal Deposition Probe for lab-scale fouling investigations

This versatile test rig consists of an extensively modified DTF, equipped with a flat-flame, multi-stage, premixed gas burner, into which the investigated solid pulverised fuel is injected. This provides adequate heating rates (10^5 K/s), well in range with full-scale PF boilers. The char particles are then led into an electrically heated reactor tube, where they are further combusted. The reactor is equipped with a conical inlet, which causes the flue gas and char/ash particles to decelerate, enabling long residence times in spite of a relative short length. Sampling equipment includes oil-cooled, quenched particulate probes, which can be coupled with a flat filter, cyclone or a cascade impactor. Furthermore, the installation is equipped with an on-line flue gas monitoring system (O_2, CO_x, C_xH_y, NO_x and SO_2). Slagging/fouling investigations can be performed by means of dedicated probes, including a recently-developed system with an on-line heat flux sensor (Figure 9.3), placed at the bottom part of the reactor, simulating superheater tubes surface.

9.4.2 DTF Application for Fuel Characterisation – Examples

The LCS has been utilised for a variety of investigations. In this section a few exemplified results on some key issues of fuel conversion are given.

Air staging strategies simulation. Air staged and oxyfuel combustion conditions can be realised in LCS by distributing the oxidant flows (air/oxygen) over the inner burner (equivalent of primary air), the outer burner (secondary air), the shield gas ring (tertiary air) as well as the Over-Fire Air facility. An example of the achievable conditions [29] is given in Figure 9.4.

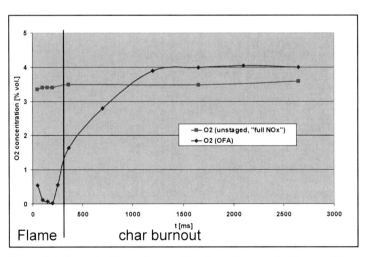

Figure 9.4 Oxygen/residence time profiles for various combustion conditions simulation in the LCS

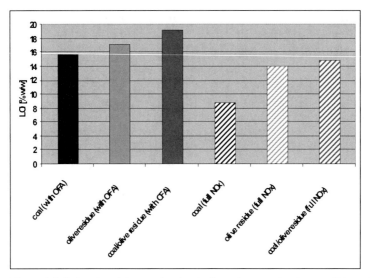

Figure 9.5 Carbon in ash/conversion levels for various combustion conditions simulation in the LCS

Conversion kinetics. Conversion kinetics are studied by sampling char/ash at various residence times or under varying combustion conditions (*i.e.* as outlined Figure 9.4) as exemplified in Figure 9.5.

NO$_x$ Formation and Evolution. Under the earlier described conditions, the formation and evolution of NO$_x$ under biomass co-firing vs air staging strategies has been studied. Results are briefly summarized in Figure 9.6.

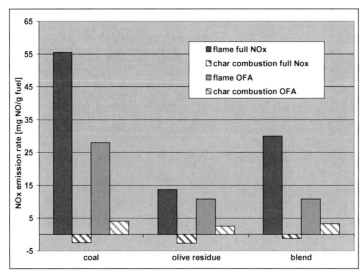

Figure 9.6 NO$_x$ formation emission rates measured in the LCS

9.5 In-boiler Diagnostic Tools

Based on the same working principle as the LCS slagging probe and its general design, an industrial-scale mobile horizontal deposition probe (Figure 9.7) has been developed for application during measurement campaigns at power plants. Unlike the lab-scale instrument, this mobile deposition probe, 3.5 m in length and 6 cm in diameter, is cooled primarily by water and only the tip of the probe, in which the sensor ring and the deposit ring are located, is air-cooled. The water cooling unit as well as the controls are fully self-contained, and require basically only electric power supply and access to pressurized air. Should this media not be available on-site, a small generator and a compressor can be used as well. Although much more complex and bulky than the lab-scale unit, the industrial deposition probe can be inserted into the boiler via manholes, scaffold supports or regular inspection ports by means of tailor-made adapter plates.

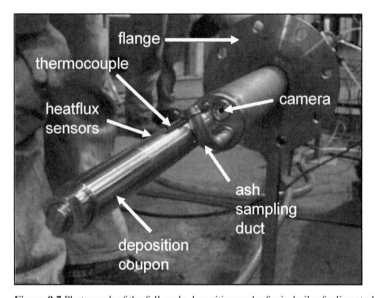

Figure 9.7 Photograph of the full-scale deposition probe for in-boiler fouling studies

9.6 Investigations of Ash Formation and Deposition under the Conditions of Biomass Co-firing in Current and Future PF Boilers

In this section, several examples of the application of the earlier described fuel characterisation techniques are given. Each paragraph presents methodology and results of lab- or full-scale investigations of one particular aspect of ash formation and deposition from relevant fuels under the conditions typical for PF boilers. Additionally, the influence of the conditions expected for the emerging technologies (*e.g.* oxyfuel firing) is discussed.

9.6.1 Ash Release

This paragraph presents results of an experimental study of the formation of ash under PF conditions. A broad range of fuels has been used, representative for current and anticipated near-future co-firing activities. Included were spruce bark and wood chips, waste wood, sawdust, olive residue, (wheat) straw and two coals from the United Kingdom and Poland.

The aim was to quantify the release of inorganic matter from the fuel. Release into the gas phase is considered to be an enabling step for interactions between the main and the secondary fuel via chemical processes such as salt formation or gas-mineral reactions, and via physical processes such as heterogeneous condensation. Understanding the basis of single-fuel ash formation and fuel–fuel interactions is a key to the successful operation of biomass co-firing plants.

Sampling and analysis. A cooled probe was used for sampling at a residence time of approximately 1,300 ms, representative of a furnace exit of a PF system. The particle residence time is taken to be that of a particle with an aerodynamic diameter of 50 μm. Residence time calculations based on the gas velocity, assuming laminar flow and taking into account the reactor geometry, axial gas temperature profile and the particle terminal velocity showed little (±10%) influence of the particle size when below 100 μm. A Pilat MARK V cascade impactor was used to obtain 11 fractions in the size range > 50 μm down to ~ 0.3 μm. Nuclepore™ polycarbonate membranes were used for their smooth surface to allow subsequent microscopic analysis. A JEOL FEG-SEM with a coupled EDX system was used to analyze each stage of the impactor. An EDX measurement was performed by scanning the whole of a deposit of particles formed underneath an orifice in the corresponding impactor jet stage.

In this way an analysis of one to three deposits per stage was obtained, including the elements Si, Al, Fe, Ca, Mg, Na, K, Ti, P, S, Cl, Mn, Zn, Pb and O. In all cases, the various analyses on a single stage were found to be very similar, indicating a homogeneous loading of the stage. Results are expressed on an oxygen-free basis. Secondary and backscatter electron images were stored for visual evaluation. From these, the particle size and morphology were determined.

Data processing. In this study the release of inorganic matter is determined as the difference between the amount of inorganic matter in the fuel and the amount of inorganic matter left over in the solid residue after (partial) combustion. Particulate matter with a particle size smaller than 1 μm – generally known as an aerial solid or aerosol – is mathematically added to the released part. The final release of inorganic matter now includes both gaseous species and aerosols, essentially referring to any inorganic matter detached from the parent fuel particles and, if particulate, with a size smaller than 1 μm. Due to the methods applied, the procedure does not discriminate between different aerosol formation routes such as evaporation–condensation or fragmentation.

For each run, the 11 impactor stages were divided into two sub-samples: A, containing the coarse char/ash, and B, containing aerosol-size particles. Samples A and B were then chemically analyzed for the elements range described earlier. For each element, the fraction released from the fuel was determined using an internal tracer. The tracer is a non-volatile (stable as a solid or liquid) element which stays with the solid phase of the fuel during combustion. The tracer was selected per fuel as one of the elements Si, Al and Fe, or a combination thereof.

Results. Figure 9.8 presents an overview of the amount of the different elements released after 1,300 ms residence time in the combustion chamber.

The data are expressed as the amount of element X released in milligrams per kilogram of dry fuel, so they can be easily applied in combustion process calculations. The percentage plotted over each stacked bar represents the mass ratio of the sum of inorganic elements released to the sum of inorganic elements in the fuel. Large differences are observed between the fuels. The release in mg/kg is influenced by the fuel ash content and the reactivity of the ash constituents. The relatively small release from the wood type fuels reflects their low ash content, while the high release from olive residue and straw is caused by the higher ash content but even more by the high ash volatility. The release from both coals is dominated by the elements sulphur and chlorine. If the sulphur content of the coal is high, the release will be high too.

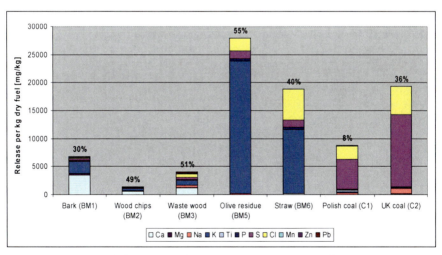

Figure 9.8 Amount and distribution of inorganic elements released after 1,300 ms residence time. Percentages represent the ratio of the sum of inorganic elements released to the sum of inorganic elements in the fuel

9.6.2   Ash Deposition – Slagging and Fouling

9.6.2.1   Biomass co-firing under state-of-the-art PF conditions.

Following the ash release study, a set of ash deposition/fouling tests has been performed, utilising the same biomass fuels, sec or blended with the Polish bituminous coal (35% biomass and 65% of coal). For the said tests, the LCS has been utilised in combination with the horizontal deposition probe. The temperature of the deposition substrate was set at 590°C, simulating superheater tubes of a state-of-the-art supercritical steam boiler.

Fouling factors. Based on the heat flux changes data measured on-line by the probe, the so-called fouling factor R_f of the obtained deposits can be estimated, which corresponds to the ash deposits heat transfer resistance:

$$\frac{1}{R_f} = \frac{1}{R_1} - \frac{1}{R_0} = \frac{T_g - T_c^1}{HF_1} - \frac{T_g - T_c^0}{HF_0} \qquad (9.1)$$

where

R_f = fouling factor, $(K \cdot m^2)/W$;
R_1 = ash deposits heat transfer resistance after time $t = t_1$, $W/(K \cdot m^2)$;
R_0 = initial heat transfer resistance after $t = t_0 = 0$, $W/(K \cdot m^2)$;
T_g = flue gas temperature, K;
T_c = coolant medium temperature inside the deposition probe, K;
HF_1 = heat flux to the sensor after time $t = t_1$, W/m^2;
HF_0 = initial heat flux to the sensor $t = t_0 = 0$, W/m^2.

By plotting the R_f vs the mass of the fuel/ash fed into the system, the specific fouling factors for each of the tests have been calculated. Results for different coal/biomass mixtures and for pure fuels are shown in Figure 9.9.

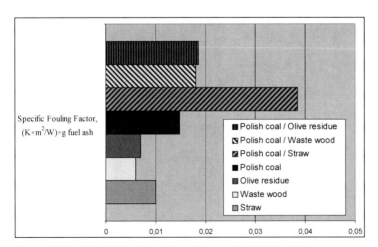

Figure 9.9 Specific fouling factor depending on the fuel/mixture burnt

As can be seen in the figure, for pure fuels combustion, the highest fouling factor was observed for Polish bituminous coal, next for straw, olive residue and waste wood. However, all of the said fuels lie in a fairly low and narrow range and only slight/moderate fouling can be expected. Under co-firing, however, each of the biomass/coal blends led to higher deposition rates in comparison with the results obtained during the individual combustion of the biomass fuels. The highest fouling factor was observed during Polish coal and straw co-firing, whereas the lowest fouling factor was measured for a mixture of Polish coal with waste wood. The co-combustion of Polish coal and olive residue gives a similar value of the fouling factor as for a blend of coal and waste wood.

9.6.2.2 Biomass Co-firing under Future PF Boiler Design Conditions

9.6.2.2.1 Slagging and Fouling in UltraSuperCritical Boiler

When using the lab-scale horizontal deposition probe and its industrial counterpart, the simulation of the future boiler designs based on the UltraSuperCritical (USC) steam conditions can be realised simply by raising the temperature of the simulated steam tube to the appropriate levels. In the study described next, the simulated steam tube temperature was set at 660°C, representing the lowest temperature of an USC superheater. The co-firing tests focusing on slagging/fouling propensities of the coal-biomass blend were performed both at lab-scale (the LCS) and at full-scale, in a Dutch 750-MWe PF power plant. The industrial horizontal deposition probe was inserted into the boiler in the area close to the primary superheater slabs, where the measured flue gas temperature was on average close to 1000°C and roughly the same as in the LCS. The fuel was a mixture of a South African coal, coffee husks and wood pellets in mass ratio of 70/15/15, respectively. Same data logging systems were applied as described in the earlier sections. After the experiment, the deposit formed on the sensor ring was collected and sent for further chemical analysis.

Fouling factors. The recorded heat flux signals for both lab- and full-scale samples are shown in Figure 9.10.

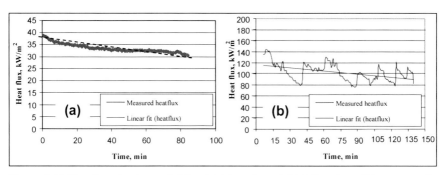

Figure 9.10 Heat flux signal recorded for: (a) lab-scale, and (b) full-scale co-firing tests under simulated USC steam tube conditions (660°C)

Figure 9.11 Photographs of the ash deposits from (a) lab-scale, and (b) full-scale co-firing tests under simulated USC steam tube conditions (660°C)

As can be seen in Figure 9.10 (b), the heat flux shows relatively big short-term variations in the full-scale measurement. Further, it can be concluded that it decreases in both experiments approximately 25% over the duration of the measurement, particularly in their first 90 min. The said fast variations are most likely due to the shearing action of the flue gas, causing part of the deposit to be removed. This can only be possible if the deposit is not very fused nor sticky. The absence of such effect in the lab-scale test is caused simply by the much lower flue gas velocity in the LCS rig.

Visual inspection. The overall low extent of the deposit formation, its powdery morphology and the absence of sintering appear to be confirmed by the visual inspection of the formed deposit, as shown in Figure 9.11.

Chemical composition. After the visual inspection, samples collected during the fouling tests have been subjected to chemical analyses. Results thereof are summarized in Table 9.1. As can be read from the data, the compositions of the collected deposits are fairly comparable, except for potassium. However, there is a simple explanation for this phenomenon.

As a consequence of the fact that the LCS is a much more dilute system (in terms of ash to flue gas ration), as compared with the full-scale plant, the condensation of alkaline metals salts may proceed to a lesser degree than in the case of the full-scale measurements. Also, during the full-scale campaign the flue gas temperature at the measurement point was some 80–100°C lower as compared with the lab-scale test, which may also affect the results somewhat.

Table 9.1 Chemical composition of the ash deposits from lab-scale and full-scale co-firing tests under simulated USC steam tube conditions (660°C)

	Lab-scale	Full-scale		Lab-scale	Full-scale
SiO_2	48.06	44.91	CaO	6.60	7.98
Al_2O_3	32.07	29.08	MgO	1.51	1.63
Fe_2O_3	7.73	7.08	TiO_2	1.58	1.69
Na_2O	0.25	0.27	P_2O_5	0.63	1.09
K_2O	1.43	6.16	MnO	0.13	0.10

9.6.2.2.2 Slagging and Fouling in Oxyfuel Combustion

Biomass co-firing under oxyfuel conditions. The combination of oxyfuel combustion with biomass and CO_2 capture is potentially the only technology forming a net CO_2 sink. However, the different gas environment experienced by the fuel particles under oxy-firing compared to conventional PF conditions, impacts the combustion processes such as ignition, elements release, char reactions and pollutant formation [30]. The effect of oxyfuel combustion on elements release and fly ash size distribution is uncertain, but it can be expected that the behaviour of certain minerals will be deeply affected. However, so far only few investigations have been performed on ash-related issues in oxyfuel combustion, even less including biomass co-firing. Krishnamoorthy and Veranth [31] modelled the combustion of a single char particle and found that varying the bulk gas composition changed the CO/CO_2 ratio within the char particle, which could affect the vaporisation of refractory oxides and in consequence the formation and composition of ash particles. The altered O_2/CO_2 ratio in the oxidant mixture has an impact on the char combustion temperature and subsequently the vaporisation of volatile elements and refractory oxides. This can also then have an effect on the composition and the concentration of submicron ash particles and aerosols. Tests in an entrained flow reactor to study the ignition and burnout of coals and blends with biomass under oxy-fuel and air (reference) conditions [32] showed varying ignition temperatures in CO_2/O_2 mixtures for oxygen concentration of 21–30% or higher. The main results of this work indicate that coal burnout can be improved by blending biomass in CO_2/O_2 mixtures. An extensive review on the available literature is presented in [33].

There are open questions prior to utilizing biomass under oxyfuel conditions, such as, for example, the ignition of blends under oxyfuel conditions, accepted gas and flame temperature profiles with respect to biomass ash components, and the degree of flue gas cleaning prior to recirculation. The increased S and Cl in the gas phase, in case they are not removed prior to recirculation, will aggravate the ash formation and deposition, inducing ash melting and affecting the fine ash particle formation. Chlorine facilitates elements volatilisation, mainly K, Na, Zn and Cd, which form chlorides increasing thus the fine particles concentration.

Oxyfuel co-firing tests in the LCS. Based on the extensive experience obtained from the LCS, further tests were performed focusing on and comparing the ash behaviour of various coal/biomass blends under air (reference) and oxyfuel conditions, in the frame of European projects. In this section, the deposition behaviour of a Russian and a South African coal with their blends with Shea meal (cocoa residues) are presented. In order to simulate flue gas recirculation into the boiler, bottled CO_2 was mixed with pure O_2. The temperature profiles in both conditions were matched by tuning the CO_2/O_2 ratios in the flame (inner burner), and further O_2 or CO_2 corrections were done in the outer burner. The horizontal deposition probe was set at 660°C to include high efficient (*e.g.* USC) boiler conditions. Ash samples from the horizontal probe as well as filter ash samples were collected.

Ash deposition ratio and deposition propensity. All of the obtained deposits were loose and powdery, indicating no melt formation. However, there was clearly a difference in the thickness between the air and oxyfuel tests – the latter deposits being significantly thicker. In order to evaluate the deposition tendency of the blends, the ash deposition ratio *DR* (9.2) was calculated, defined as the ratio of the ash collected on the deposit probe during the experiment, divided by fuel fed for the same time frame. In order to normalise the ash deposition in relation with the fuel ash content, the deposition propensity *DP* (9.3) is introduced, defined as the percentage of the ash on the deposit to the ash fed by the fuel. The deposition propensity, here expressed in %, provides more insight into the inherent deposition characteristics of the different fuels as it accounts for variations in fuel ash content:

$$DR = \frac{m_{dep}}{m_{fuel}} \tag{9.2}$$

$$DP = \frac{m_{dep}}{m_{ash}} \times 100\% \tag{9.3}$$

where:

DR = Deposition Ratio (–);
DP = Deposition Propensity (%);
m_{dep} = mass of the deposited ash on the probe substrate during a deposition test (g);
m_{fuel} = mass of the fuel fed into the LCS during a deposition test (g);
m_{ash} = mass of the ash fed though the fuel into the LCS during a deposition test (g).

Both parameters are shown in Figure 9.12.

In order to understand the phenomena causing the observed experimental differences one must consider the deposition mechanisms, governed by the processes such as the inertial impaction including impaction and sticking, thermophoresis, condensation and the chemical reactions of the deposit and gas-bound ash particles [1]. First, the inertial impaction, prevailing in reactors as the present one, is dependent on the variations in the physical gas properties, such as, *e.g.* the gas density of CO_2/O_2 mixture, which is higher than under N_2/O_2 conditions. This may in part explain the higher deposition rates under oxyfuel firing. Second, the flame temperature profile difference between CO_2/O_2 and air firing may also be responsible. Even though the flame temperatures in both series of tests were very close, a difference in the flame length was observed, as the oxyfuel flame appeared to be longer and with a more diffused flame front. In practice, this translates into slightly longer residence times of the particles in the peak flame zone of the oxyfuel flame, which in turn could lead to melt formation and ash particles that are more prone to deposition [34].

Ash partitioning. In order to compare directly the distribution between the deposited ash and the ash collected in the filter under the two combustion environments, these two ash quantities are expressed as ash distribution percentages of the total ash fed (Figure 9.13).

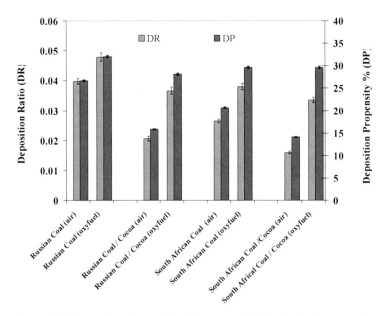

Figure 9.12 Deposition ratio and deposition propensity for the fuel blends combusted in O_2/CO_2

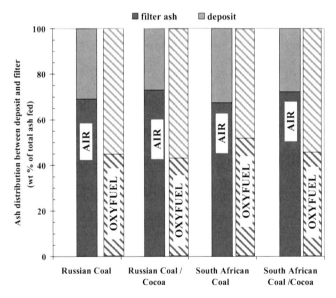

Figure 9.13 Ash distribution between deposit and filter ash for the fuels and blends combusted in air and O_2/CO_2

In Figure 9.13 it can be observed that the filter ash yields under air firing conditions are systematically higher. With respect to the lower deposition results during air combustion, the same fuel mass flow rates and fuel quantities, and excluding the presence of a molten slag formation in all the deposit probes, one could eventually support the conclusion that the different combustion environment (CO_2/O_2) affects the ash deposition behaviour when shifting from air to oxyfuel combustion.

More research and targeted lab-scale tests are needed in order to validate a wide variety of biomass fuels co-fired under oxyfuel conditions, prior to advancing this technology.

References

1. Baxter LL (1993) Biomass Bioenerg 4(2):85–102
2. Sarofim AF, Helbe JJ (1993) In: Williamson J, Wigley F (eds) The impact of ash deposition on coal fired plants. Proceedings of the Engineering Foundation Conference, June 20–25, Solihull, England. Taylor & Francis London, 1994
3. Sarofim AF, Howard JB, Padia AS (1977) Combust Sci Technol 16:187–204
4. Lith van SC (2005) Release of inorganic elements during wood-firing on grate. PhD Thesis, CHEC Research Centre, Technical University of Denmark, Lingby, Denmark
5. Schurmann H, Monkhouse PB, Unterberger S, Hein KRG (2007) Proc Combust Inst 31:1913–1920
6. Kramlich JC, Chenvert B, Park J (1995) DOE grant no. DE-FG22-92PC92548, Quarterly Technical progress report no. 10 for U.S Department of Energy
7. Kramlich JC, Chenvert B, Park J (1995) DOE grant no. DE-FG22-92PC92548, Quarterly Technical progress report no.11 For U.S Department of Energy
8. Baxter LL (1992) Combust Flame 90:174–184
9. Buhre BJP, Hinkley JT, Gupta RP, Wall TF, Nelson PF (2005) Fuel 84:1206–1214
10. Thy P, Lesher CE, Jenkins BM (2000) Fuel 79:693–700
11. Lind T (1999) Ash formation in circulating fluidized bed combustion of coal and solid biomass. PhD thesis. Helsinki University of Technology, Helsinki, Finland
12. Wigley F, Williamson J, Gibb WH (1997) Fuel 76(13):1283–1288
13. Dacombe P, Pourkashanian M, Williams A, Yap L (1999) Fuel 78:1847–1857
14. Yan L, Gupta RP, Wall TF (2001) Fuel 80:1333–1340
15. Brink ten HM, Eenkhorn S, Weeda M (1996) Fuel Process Technol 47:2233–2243
16. Yan L, Gupta RP, Wall TF (2001) Energ Fuel 15:389–394
17. Baxter LL, Lu H, Ip E, Scott J, Foster P, Vickers M (2008) Fuel prepublished on the web
18. Mathews JP, Hatcher PG, Scaroni AW (1997) Fuel 76:359–362
19. No SY, Syred N (1990) J Inst Energ 63:151–159
20. Korbee R, Cieplik MK (2006) Ash formation from biofuels and coal in pf systems. In: Impacts of fuel quality on power production. Snowbird, UT, USA, October 29–November 3
21. Frandsen FJ, Lith van SC, Korbee RK, Yrjas P, Backman R, Obernberger I, Brunner T, Joeller M (2006) Quantification of the release of inorganic elements from biofuels. In: Impacts of fuel quality on power production. Snowbird, UT, USA, October 29–November 3
22. Slot van der HA, Meeussen JC, van Zomeren A, Kosson DS (2006) J Geochem Explor 88:72–76
23. van Eijk RJ, Stam AF, Cieplik MK (2009) The 34th International Technical Conference on Coal Utilization and Fuel Systems, Clearwater (Fl), USA, May 31–June10
24. Coda B, Cieplik MK, Wild de PJ, Kiel JHA (2007) Energ Fuel 21:3644–3652
25. Bertrand CI, Losurdo M, Korbee R, Cieplik MK, van de Kamp WL (2007) In: Biomass co-firing in boiler and furnaces: progress in the development of an ash deposition post-processor. ECCOMAS Conference on Computational Combustion, Delft, the Netherlands, July 18–20
26. Improved coal utilisation strategies by standardisation and wider use of Drop Tube Furnace evaluation methods. EU Research Fund for Coal and Steel project ECSC 7220-PR-106

27. Spliethoff H (2009) Investigations on high temperature gasification and gas cleaning – the research project HotVeGas, In: 4th International Conference on Clean Coal Technologies, Dresden, Germany, May 18–21
28. Korbee R, Boersma AR, Cieplik MK, Heere PGTh, Slort DJ, Kiel JHA (2003) ECN contribution to EU project ENK5-1999-00004 Combustion Behaviour of 'Clean' Fuels in Power Generation (BioFlam)' ECN report ECN-C-03-057
29. Cieplik MK, Verhoeff F, Korbee R, Kalivodová J, Kamp van de WL (2008) Final Technical Report by ECN within the EU-MINORTOP Project' ECN report ECN-E-07-078
30. Sheng C, Li Y, Liu X, Yao H, Xu M (2007) Fuel Process Technol 88:1021–1028
31. Krishnamoorthy G, Veranth JM (2003) Energ Fuel 17:1367–1371
32. Arias B, Pevida C, Rubiera F, Pis JJ (2008) Fuel 87:2753–2759
33. Fryda L, Sobrino C, Cieplik M, van de Kamp WL (2010) Fuel 89(8):1889–1902
34. Wigley F, Williamson J, Riley G (2007) Fuel Process Technol 88:1010–1016

Chapter 10
Utilization of Biomass Ashes

J.R. Pels and A.J. Sarabèr

Abstract Useful application of ashes produced in the thermal conversion of bio-mass can contribute to the green image of biomass as a source of sustainable energy. This chapter gives an overview of the different forms of ash utilization that exist or are being developed for biomass ashes. The first section reviews options for ashes from co-firing of biomass and coal, both established forms of utilization in cement and concrete, and alternative options, *e.g.*, manufacture of lightweight aggregates. The second section discusses utilization options for residues from "pure" biomass combustion. The large variation in biomass fuels and installation types makes this a complex issue. Besides recycling of clean wood ash to forests, these are all emerging forms of utilization. The third section discusses the specific issues related to the utilization of carbon-rich ashes from biomass gasification and pyrolysis.

10.1 Introduction

Thermal conversion of biomass fuels results in various residues. These are various ashes, spent bed sand from circulating fluidized beds, and residues from flue gas cleaning, *e.g.*, gypsum. In order to preserve the green image of biomass as a source of sustainable energy, these residues should find useful applications.

Ash utilization is often neglected when thermal installations for energy production from biomass are installed. In many cases, the issue of ash utilization is postponed until the installation is up and running. This makes sense for two viable reasons:

J.R. Pels (✉)
Energy research Centre of the Netherlands (ECN),
P.O. Box 1, 1755 ZG Petten, The Netherlands, Tel: +31 224 56464884,
e-mail: pels@ecn.nl

A.J. Sarabèr
KEMA, P.O. Box 9035, 6800 ET Arnhem, The Netherlands, Tel: +31 26 3562412,
e-mail: Angelo.Saraber@kema.com

1. Exact ash compositions are not easy to predict and depend on the fuel and the type of installation.
2. There is always the option to landfill ashes. However, the costs of landfill can be high and utilization is investigated for being a more economic option.

The composition of ashes from biomass is significantly different from coal ashes, so that established routes for coal ash cannot be used. Only for ashes from co-firing, where the coal ash dominates the ash composition, are established routes for coal ash utilization an option. In "pure" biomass ashes, the high contents of chloride, alkali metals, phosphorus, and calcium make them unsuitable for the established applications for coal ashes. In co-firing at high levels, the residues from thermal conversion of biomass also do not fulfill the requirements of utilization options existing for coal ashes.

Several studies of ECN, KEMA, and others [1–5] have provided overviews of residue properties depending on the type of conversion process and biomass properties. These studies also give indications with respect to possible recycling and utilization options. The general conclusions from these studies are presented in this chapter.

10.1.1   Types of Ashes

In principle, for utilization of ashes it is not necessary to know where the ashes were collected and how they were formed in the installation. The only things that matter are the physical and chemical characteristics of the waste streams from thermal conversion installations that use biomass as a fuel. For the sake of simplicity, in this chapter ashes are put into only two groups:

- bottom ashes, including slag, dry bottom ashes and bed solids from fluidized beds;
- fly ashes, including boiler ashes, cooler ashes, cyclone ashes, and filter ashes.

The basic difference between bottom ashes and fly ashes is that fly ashes are entrained by the flue gas and are subsequently separated from the flue gas, while bottom ashes drop from the flames and are collected at the bottom of the thermal reactor. The general difference in characteristics between bottom ashes and fly ashes makes a first screening for the most appropriate utilization options possible.

Most bottom ashes from thermal conversion of biomass and from co-firing of biomass with coal have common characteristics, more or less independent of the fuel and – to a certain extent – reactor type:

- The bulk of bottom ashes is formed by sand and sand-like, inert materials.
- Bottom ashes have a low loss-on-ignition (LOI) and almost no volatile compounds.
- Bottom ashes have a low or zero calorific value.

- There are low concentrations of nutrients in bottom ashes. When phosphorus is present it is typically in an inert form.
- Bottom ashes have a low concentration of leachable contaminants.

These common characteristics even apply to bottom ashes from gasification. It makes bottom ashes in general attractive for direct utilization as building material, notably in road construction and other infrastructural works.

Fly ashes from thermal conversion of biomass have the following common characteristics:

- Fly ashes are fine powders, with very low bulk density.
- Fly ashes contain ash-forming components from the biomass fuel, and in the case of fluidized beds, are supplemented with fractured bed material.
- Fly ashes have large variations in composition.
- Fly ashes exist with very low and very high LOI, carbon content, and calorific value, depending on the biomass type, installation type, and operation conditions.
- Fly ashes contain those elements that are volatilized during combustion and condensed when the flue gas is cooling.
- Fly ashes typically show high leaching.

The large variation in fuel composition make it difficult to identify a generalized most likely form of utilization that can be commonly applied to fly ashes. In fact, the large variation in fly ash volumes inhibits development of utilization options.

10.1.2 Approach to Finding Utilization Options for Biomass Ashes

Ashes are by-products of heat and power production. The ashes and their composition are the starting point for a search to find the best utilization option. For each ash, various forms of utilization of biomass ashes should be listed, screened, and ranked. First, it should be verified which options on the list are technically viable. Subsequent ranking of technically viable options can be done using the EU Strategy for Waste Management of 1989 [6] as illustrated in Figure 10.1.

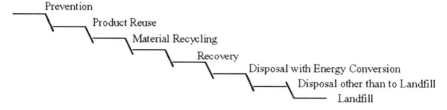

Figure 10.1 Preferences for utilization of waste materials, based on the proximity principle

The highest level is prevention. This is possible when biomass fuels without ash-forming components are used. The best example is vegetable oil. When the press cakes after oil extraction are used as animal feed, there will be no ash that needs to be utilized.

The next best is product reuse. An example is reburning of ashes in a power plant to generate power from the unburnt carbon fraction of the ash.

Recycling is when the ashes – in fact the nutrients – are returned to the soil where biomass fuels originated.

Recovery is a broad category and includes both direct applications and utilization as a raw material in the manufacture of products. These applications can be with or without pre-treatment and form the majority of the utilization options that are pursued for biomass ashes. Well-known examples are the use of fly ash in concrete and bottom ash in road construction. Extracting phosphate from ashes for fertilizer production is also within the category recovery.

Disposal with energy recovery takes place when ashes go to a waste incinerator, which also produces heat and/or power.

Disposal other than landfill may be application in artificial reefs.

Landfill is the least desirable option, but it is common practice for many small and poorly described ash volumes. Landfill is not the same as landscaping[1].

10.1.3 Bulk and Niche Applications

Ashes from thermal conversion of biomass are produced in large quantities. Therefore it is necessary to put priority on seeking bulk applications. Bulk applications of fly ash can be classified into three general categories:

- building materials and building products;
- fertilizer;
- fuels (only for carbon-rich fly ash).

Within these general categories many different forms of applications exist. The category of building materials includes directly utilization, *e.g.*, in road construction, and utilization as raw material in the manufacture of building products, *e.g.*, as filler in concrete and cement, or in lightweight aggregates. The same applies to fertilizers: direct utilization is possible[2], as well as utilization as a raw material in the manufacture of other fertilizers. Ash recycling to forests is a special form of utilization as fertilizer.

Niche applications may be attractive and may be used for image building, but they are not going to solve the question of where to put the large ash volumes

---

[1] The difference between landfill and landscaping is that landfill would not have been done if there were no ashes that need disposal. Instead, another material, *e.g.*, sand, is used if there are no ashes available. Thus, landscaping can be regarded as a form of utilization as building material.

[2] Using ashes directly as a powder is perhaps not a good idea. Wetting and granulation prevents dusting in the application.

from combustion and gasification. Niche applications can have high economic margins, which make it possible to perform some kind of after-treatment, *e.g.*, thermal treatment or washing. For bulk applications, after-treatment should be avoided, except for very low cost operations like screening.

10.2 Ashes from Co-firing Biomass with Coal

Pulverized coal combustion (PF combustion) is one of the main technologies for power generation from coal that is suitable for co-firing of biomass. The coal is pulverized and combusted in burners in a boiler. When biomass is co-fired, it is mostly done directly by means of burning pulverized biomass in the burners, mixed with coal or combusted in separate burners. As a result, the ash forming components in the biomass are mixed with those from coal and mixed ashes are formed. The bulk of the ashes are entrained with the flue gas. In electrostatic precipitators (ESP), fly ashes are separated from the flue gas. In PF combustion about 10–15% of the ash is retrieved as bottom ash and the rest as fly ash [7].

In the case of indirect co-firing, biomass is gasified in a separate installation. Before the producer gas is fed to the boiler, it is cleaned, *e.g.*, using a cyclone for dust removal. In this way, it is possible to maximize the amount of biomass that can be co-fired, and to minimize the impact on the main boiler. Indirect co-firing is the preferred option for biomass fuels that have high amounts of unwanted components, such as potassium chlorine (*e.g.*, agro-residues) or heavy metals (demolition wood). The AMER-9 power plant in the Netherlands is a good example where most of the ash-forming components present in the wood fuel are separated, but some – in particular those that are volatilized – go to the main boiler and are eventually mixed with the coal ash. Indirect co-firing results in separate ash volumes that are essentially ashes from gasification of biomass only and can be treated as such (see other sections of this chapter).

The commonly used co-firing rates are rather low, typically 5–10% on an energy basis. Since most biomass has ash contents that are significantly lower than coals, this means that ashes from co-firing are dominated by coal ashes. The effective contribution to the ashes from biomass is just a few percent. Therefore, the forms of utilization normally used for coal ashes can still be used[3].

10.2.1 Co-firing Bottom Ashes

The most common form of utilization for bottom ashes is in road construction and other infrastructural works, where it is used in the embankment and foundations.

---

[3] Maintaining the existing utilization routes for coal ashes is one of the factors that determine the maximum for co-firing rate. Security of supply, emission limits, slagging, and corrosion risks are other important factors.

The key in this application is whether contamination through leaching of unwanted elements stays below the legal limits. In the Netherlands, the Dutch Building Materials Decree (DBMD) [8] was used to determine whether materials could be applied and most of the applications in this chapter are tested using the criteria of the DBMD. Recently, it has been replaced with the Soil Quality Decree [9]. Both regulations are based on standardized leaching tests (CEN/EN-12457, *etc.*) and model calculations. In the near future, the European Construction Products Directive and its amendments will apply, replacing national regulations.

In practice, most bottom ashes from coal-fired plants, with or without co-firing of biomass, can be used with little or no restrictions. Sometimes the ashes need aging (exposure to ambient air) so that they adsorb carbon dioxide and become less basic, which diminishes leaching of certain elements. The standard CEN/EN-13055, provides requirements for the use of bottom ash in civil engineering.

10.2.2 Co-firing Fly Ashes in Concrete and Cement

For fly ashes from coal-fired plants, the most common form of utilization is as filler in concrete. The standard CEN/EN-450 provides a set of requirements to assess the quality of fly ash for this application. In concrete, fly ash is a valuable addition due to its pozzolanic behavior, which provides concrete higher durability. Another important application of fly ash is in cement and cement products. Also, for these applications, standards exist, like CEN/EN-197. For applications with a limited number of end-users, requirements are not established in standards but in contracts, like the use of fly ash as raw material for the production of Portland clinker.

Nearly all fly ashes from coal combustion in modern pulverized fuel (PF) boilers match the requirements for CEN/EN-450. When biomass is co-fired, the fly ashes may not comply anymore. In general these ashes are enriched in free CaO (2–3 wt%), K_2O (4– wt%) and/or P_2O_5 (3– wt%). The ashes consist mainly of glass and to a lesser extent quartz, hematite/magnetite, and phosphates. The LOI (loss on ignition) may be in the range 3– wt%. Co-combustion (or co-gasification) of contaminated biomass like demolition wood may cause an increase of the content of heavy metals in the ashes like Cu, Pb, and Zn.

Depending on the exact biomass and coal characteristics, the maximum co-firing rate varies greatly. In the case of clean wood and typical coals for PF boilers, the co-firing rate can be as high as 25–40% (energy basis). Currently, fly ashes from coal-fired plants that co-fire biomass are within the limits of the CEN/EN-450. For power plants there are good reasons to comply with those limits, because landfill of fly ash is considerably more expensive than utilization in concrete or cement. Of course, this strongly depends on local conditions.

In the future, when very high co-firing rates may be achieved, co-firing ashes will be more like mixed coal/biomass ashes. For bottom ashes, few problems can be expected. The challenge will be to find new applications for fly ashes with a chemical composition that no longer matches with the requirements of the CEN/EN-450[4]. At the moment, there are several options that are technically feasible and can handle large volumes. However, none of these are economically attractive. A recent overview by KEMA and ECN [4] indicates that production of lightweight aggregates (LWA) is closest to economic success.

10.2.3 Co-firing Fly Ashes in Lightweight Aggregates (LWA)

Typical process steps for the manufacturing of LWA are the agglomeration and the bonding of the aggregate particles. Agglomeration techniques are subdivided into agitation and compaction, bonding methods into melting, sintering, and cold bonding. At the moment, only processes based on cold bonding and sintering with coal fly ash are implemented[5]. Cold bonding is based on matrix formation by calcium silicate hydrates obtained by cement reactions and/or pozzolanic reactions between the fly ash and lime. The reactions take place at room temperature up to 250°C (Aardelite process).

Sintering is based on partial melting, which takes place in the range 1,000–1,200°C (e.g., Lytag process). There are two basic principles: sintering with a rotary kiln and with a sintering band. Rotary kilns are being used for LWA produced with clays (ARGEX) but also with secondary materials (Trefoil process). Sintering bands are used for the production of Lytag LWA.

10.2.4 Other Options for Co-firing Fly Ashes

A wide range of other potential bulk and niche applications for co-firing fly ashes has been collected and assessed [4]. Besides LWA, two other applications were identified as being forms of utilization that can be used for the bulk of co-firing fly ashes, namely application in infrastructural works and back-fill of mining. Decreasing mining activities and competition with other waste streams make these options economically less attractive.

For most other options that are technically feasible, the market may be too small in order to take the large volumes of fly ash that are produced by co-fired power plants (Figure 10.2). This means that the fly ash from one or a few power plants can be used. Examples are manufacture of sand-lime bricks, special cement mixes, or mineral fibers.

---

[4] Adjusting the limits of CEN/EN-450, which has been done in the past, will no longer be the solution when concrete quality is jeopardized.
[5] Melting processes are implemented for other waste streams, e.g., municipal waste incineration. Melting processes are based on total melting in the range 1,500–1,600°C. The energy consumption, needed to reach these temperatures makes melting processes economically not attractive.

Figure 10.2 Picture of LWA from mixed coal/biomass fly ash

10.3   Ashes from Combustion of Biomass Only

The ashes from installations where only biomass is combusted vary widely in composition. First, there is a large diversity in fuels, including clean wood, straw, and coffee husks. Second, different types of installations are used, *e.g.*, grate stokers, fluidized beds, and even specially designed burners for bales of straw. This large variation in ashes and ash qualities makes finding utilization options a complex issue. Only for a few types of ash have applications been established. Developing new technology is nearly impossible because of the high investments needed. As a consequence, many ashes are landfilled.

Whenever possible, biomass ashes should be used in existing routes. Bottom ashes from biomass combustion are likely to have a composition similar to bottom ashes of coal combustion or waste incineration. As mentioned previously, these are often used in road construction and other infrastructural works. Therefore, these routes can be explored successfully. Again, the most difficult ashes are the fly ashes.

Due to the large variation of ash characteristics from "pure" biomass combustion, it is not possible to identify a single most attractive form of utilization. In the sections below, nutrient recycling, utilization as fertilizer, and several options in building materials are discussed.

10.3.1   Nutrient Recycling

During growth, plants take up nutrients from the soil. When harvested, plants and nutrients are removed from the area where the biomass grew. For a sustainable operation, it is necessary to restore the original level of nutrients; otherwise the soil can be depleted of minerals. Since after combustion of biomass many of these nutrients end up in ashes, it is logical to return biomass ashes to the land where the biomass grew.

In forestry and energy plantations, true recycling is possible. Nearly all nutrients that were harvested along with the trees are concentrated in the ashes from known combustion installations. Often these are the heat and power plants of the local wood processing industry. These ashes have a well-known origin and can be used to compensate for the loss of nutrients in the forest soil.

In agriculture, recycling to the original soil is difficult. The biomass used for combustion contains only a fraction of the nutrients; the rest is in the food that is produced or is fed to animals. Also, it is nearly impossible to trace back the original soil of agro-residues. Thus, at best, biomass ashes can be used as a fertilizer together with other fertilizers and manure. This form of utilization is classified as use as fertilizer, not recycling.

10.3.2 Nutrients in Biomass Ashes

The suitability of biomass ashes as a source of nutrients for plants depends strongly on the actual composition. Ash from clean wood pellets contains mostly Ca and Mg, but the ashes from agro-residues and manure can be rich in phosphorus (P) and potassium (K), two of the three main nutrients in agriculture. Nitrogen (N) is missing, since it is emitted in the flue gas after combustion. Other important nutrients that can be present in biomass ashes are sulfur (S), calcium (Ca), magnesium (Mg) and dozens of trance elements.

When biomass ashes are put directly in soils, K is easily leached and immediately available for plants. P is often in an insoluble form (e.g., apatite) and requires up to 20 years before it has become fully available for plants. In forestry this is no problem, because the growth cycle of trees is 20–60 years, depending on species and climate. In agriculture, the low solubility may prevent direct utilization as P-fertilizer. For rapid plant availability, the ashes need to be dissolved in strong acid and then further processed, just like phosphate ores. Therefore, biomass ashes can better be used as a raw material for fertilizer manufacture or for long-term purposes.

Biomass ashes containing high amounts of potassium and phosphorus, e.g., from combustion of forest residue, wood trimmings, fast growing biomass plantations, and certain agro-residues are most suitable for recycling and use in fertilizers. Biomass ashes containing primarily calcium and magnesium, e.g., from combustion of bark, can be used as soil improver to balance pH.

It should be noted that these "pure" biomass ashes may contain sand and soil. In the case of fluidized-bed combustion, bed sand and additives like lime can also be found in the ashes. Normally, these added minerals do not prohibit recycling or the use as fertilizer.

10.3.3   Recycling of Ashes in Forestry

The nutrients in trees are concentrated in the living parts of trees, the leaves, needles, twigs and bark, but not in the stem wood. When trees are harvested, traditionally, the branches and tree tops remain at the harvest site as forest residues. This form of harvesting is called Stem Extraction (SE). There they slowly decompose and the minerals return to the soil. The stem wood contains almost no ash-forming components. Erosion and deposition compensate for the loss of these minerals[6].

In the last decades it has become lucrative to collect forest residues and use it as cheap, clean fuel. The forest residues are collected and bundled[7]. This form of harvesting is called Whole Tree Harvesting (WTH) and has a significant risk of depleting the forest of nutrients. In Sweden, WTH is only allowed when ashes are recycled. The Swedish Forestry Agency (Skogsstyrelsen) has set up guidelines and recommendations together with the industry for ash recycling [11, 12]. These include dosage at specific years after harvesting and quality standards for the ash. Ashes suitable for recycling are collected from combustion installations that burn only clean wood. The ashes then need to be stabilized before spreading, to avoid fast leaching. It is not necessary to recycle ashes to exactly the same location as where the trees were harvested; however, ashes should have a composition similar to the ashes that would remain from combustion of the trees from that location.

The application of biomass ashes in forestry should never interfere with general legislation on environmental (soil) protection, in particular regarding heavy metals. Plants may have grown on polluted sites and ash recycling should not perpetuate pollution. Also, speciation of elements may change, creating toxic constituents that were not present in the soil or the plant, *e.g.*, oxidation of Cr(III) to Cr(VI).

A complete overview of wood ash recycling can be found in the *International Handbook – From extraction of forest fuels to ash recycling* [13].

The above considerations regarding WTH apply also to energy plantations where fast growing trees like poplar and willow are grown, as well as crops like bamboo or miscanthus. As long as the nutrient-rich parts of the plants are combusted in known installations, the potassium and phosphorus containing (fly) ashes from those installations can be used for recycling to the soils of the energy plantations.

---

[6] It depends on local conditions whether the removal of bark together with the stem wood is significant. Bark contains Ca and Mg. These are not elements that limit the growth of the forest, but removal of Ca and Mg through removal of bark at the felling site may cause slow acidification in sensitive soils.

[7] The bundles can be removed directly, Green Harvesting, or left for several months on site or along the road, Brown Harvesting. The latter is done to let leaves and needles drop out and to return nutrients to the soil. However, when this is done along the road, the felling site where the trees grow is not benefiting. Also, research presented at the RECASH workshops in Prague and Karlstadt [10] has shown that 70% of the nutrients still remains in the forest residue in the case of Brown Harvesting.

Finally, it should be noted that in more densely populated regions, forests are primarily used for recreational purposes. There, ash recycling may not be feasible because it is undesirable to bring the public into contact with ashes.

10.3.4 Fertilizer Use

The applicability of combustion fly ashes from clean wood in the Netherlands has been assessed [14]. In the Netherlands, when a material is used as a fertilizer, it must be listed on the list of approved fertilizers [15]. Applicability of a material to be used as a fertilizer is based on assessment of nutrients content *vs* contaminants.

The nutrient value of a material is determined from the total or available content of N, P, K, Ca, S, and other elements, resulting in a minimum required load (in kg per hectare per year). The environmental impact is calculated from the total contents of Cd, Cr, Cu Hg, Ni, Pb, Zn, and As. Maximum limits are taken from the Dutch regulations for agricultural use of sewage sludge (the Dutch Decree of Other Organic Fertilizers), resulting in a maximum allowed load. Only when the maximum allowable load exceeds the minimum required load a material can be allowed as a fertilizer.

The best opportunities for application as fertilizer and/or phosphor production involve ashes from combustion of specific agricultural residues such as cacao residues, olive shells, coffee husks, *etc.* At the moment, ashes from combustion of chicken litter[8] are being utilized as fertilizer (*e.g.*, from Fibrowatt plant in the United Kingdom) or as raw material for fertilizer production (*e.g.*, from BMC Moerdijk in the Netherlands). Furthermore, meat and bone meal could be suitable as these ashes have high contents of K, P, and Ca.

Certain fly ashes from combustion of clean wood were found to contain significant amounts of the nutritious elements Ca (varying between 12 and 15%) and K (over 30%). Other elements cannot be regarded as nutrients since their concentrations were too low. Local fertilizer regulations will determine whether biomass ash from combustion of wood can be on the list of approved fertilizers. For the Netherlands, wood ashes investigated [2] contain contaminants in such concentrations that the maximum allowed load was in all cases too low for useful application[9]. In practice, the Cd content is the limiting factor.

---

[8] Strictly speaking, chicken litter is manure, which is not a biomass fuel, but a waste material. Combustion of manure is regulated by the EU Waste Incineration Directive.
[9] The maximum allowed load is the amount of fertilizer that can be applied per hectare of farmland. The minimum required load is the amount of fertilizer that is required to be effective for next year's crop. In The Netherlands, fertilizers are only allowed when the minimum required load is less than the maximum allowed load [14].

10.3.5   Use in Building Products

The application of ashes from "pure" biomass ashes in building materials is still underdeveloped. Often high concentrations of potassium and chloride make these applications unattractive due to leaching and corrosion risks. Also, quite unpredictable levels of heavy metals make these ashes less attractive. A study by ECN and KEMA has identified a few options that may be worth further investigations.

The ashes from various kinds of biomass may be used for synthetic aggregates (see section on co-firing ashes) or soil stabilization and landscaping. Another application is the use of biomass ashes from combustion of sewage sludge and demolition wood into cement-bound or bitumen-bound road construction materials. Several examples were presented at AshTech 2006 [16].

One of the more exotic options is the use of biomass ashes in concrete used in artificial reefs (Figure 10.3). There, leaching of alkali and chloride is of no concern. The concrete blocks have no heavy structural function. Over time, the blocks will overgrow or disintegrate, which makes the depositing of new blocks necessary. The use of coal fly ash in artificial reefs blocks have been investigated by several institutes in Japan, United Kingdom, and the United States. Research data from the UK shows no significant differences between the heavy metal content of organisms growing on the ash reef blocks and those on the concrete reef. The possibility of transfer and concentration of excess heavy metals by predatory fauna higher in the food chain has been considered and no evidence was found for transfer of heavy metals from the reef blocks to the epifauna. These observations were confirmed by other observations of artificial reef projects with fly ash and other secondary materials like the artificial reefs project in the Tyrrhenian Sea near Torre Valdaliga, Italy [17].

Figure 10.3 Artificial reefs (Reefballs) made with precast concrete near Dreischor in Zeeland ([©] Reefball International) [18]

10.3.6 Consistency and Niche Applications

At the moment, none of the solutions presented above for "pure" biomass is large enough to be identified as bulk application. Recycling to forests may be the first to outgrow the niche status. For all applications, but in particular for niche applications, consistency of ash quality is of key importance. Since niche applications typically involve products with high added value, there is more economic margin to switch (back) to virgin materials[10].

10.4 Utilization Options for Carbon-rich Fly Ash

Carbon-rich fly ashes are residues from gasification and pyrolysis, generally incomplete conversion. The fly ash from gasification ash is typically composed of ash particles, mixed with non-porous carbon particles. Carbon-rich ash from pyrolysis may be porous and resemble the open fibrous structure of the original biomass, depending on the pyrolysis temperature and the mechanical stress during the process.

Carbon-rich fly ashes form a specific category of fly ashes due to a high carbon content and need a specific approach. There are no established routes for utilizing this kind of biomass ash. Applications discussed here are experimental or under (academic) investigation.

Options for bulk utilization of carbon-rich fly ash from CFB gasification of clean wood and demolition wood were thoroughly investigated for potential uses in the EU funded project GASASH [3]. Below, the discussion is centered around the carbon-rich fly ash from demolition wood using criteria that apply to the Netherlands. For other ashes and other countries, the results may be different but the general approach is the same. Another overview is presented by Gómez-Barea *et al.* [5], focusing on carbon-rich ashes from waste gasification.

10.4.1 Building Material

The opportunities to utilize carbon-rich fly ash directly as a bulk building material are strongly limited. Mainly, the physical and mechanical characteristics prevent effective use. It is a fine black powder of very low density. Some carbon-rich fly ashes may self ignite; others are inert.

---

[10] This is of even greater interest when niche applications are new and under development. Poor quality products, *e.g.*, due to inconsistent quality of the raw materials, may damage your reputation and even the development of the application itself. Disappointments can easily prompt manufactures to pay a little extra for virgin materials, just to be on the safe side.

In principle, carbon-rich fly ash may find direct application in road construction when compacted. In the case of carbon-rich fly ash from demolition wood, its utilization as bulk building material is prevented due to leaching behavior of metals being present such as Ba, Cl, Pb, Zn and Br [3]. Carbon-rich fly ash from clean wood may pass the test.

Indirect application, as a component in the manufacture of building materials, is a more likely option. An interesting option is to use carbon-rich fly ash in the production of lightweight aggregates (LWA). The carbon content of fly ash from modern coal fired power plants is below the optimal carbon content for the production of LWA. Blending with carbon-rich fly ash instead of powder coal may be used to reach the optimal carbon content for sintering and thus improving product quality.

Carbon-rich fly ash can also be used as filler in asphalt and asphalt-like products. Leaching of unwanted components is strongly inhibited because bitumen covers the all ash particles and limits contact with water. It is technically easy and works for nearly all kinds of fly ash and other solid residues. Unfortunately, when carbon-rich fly ashes must compete with residues from waste incineration, and similar prices are charged for gasification ashes, it is not an economically attractive route.

A niche application that is possible for all carbon-rich fly ashes is the utilization as filler in C-Fix blocks. C-Fix is a material building composed of blocks; it is composed of gravel, sand, filler, and bitumen as binder. Its main application is as a support layer under asphalt roads [19]. In the GASASH project, fly ash from gasification of demolition wood has been applied to produce two C-fix test blocks. Both blocks passed the tank leaching test (NEN 7375) and complied with the DBMD for 'shaped' materials. The bitumen prevents contact of water with fly ash and leaching is very low. The tests showed that C-fix blocks made with gasification fly ash fulfilled the requirements for compressive and flexural strength and can be used without restrictions [3]. The application in C-Fix is a technically feasible option, but economic feasibility is uncertain.

10.4.2 Fertilizer

In principle, carbon-rich fly ash from gasification can be used as a fertilizer. The presence of carbon is not an objection. It contains nutrients, K, Ca, and S, possibly P in ash particles. Carbon behaves as inert bulk. The carbon is neither a nutrient nor a contaminant, and thus the situation is comparable to utilization of fly ash from combustion. An additional issue that may prohibit the use of carbon-rich fly as fertilizer may be that the ash is black and may stain the soil visibly. In other countries the situation may be different, in particular in locations where the soil is significantly depleted of carbon.

In the GASASH project [3], it was quickly established that gasification ash from demolition wood contains too many contaminants compared to nutrients. For carbon-rich ash from clean wood, application as fertilizer may be possible.

Although, in general, utilization as fertilizer does not appear to be the best suited route for carbon-rich ashes, certain ashes may be suitable; *e.g.*, fly ash from gasification of chicken litter with a high burn-out and a high nutrient content may be used for the production of fertilizer.

Recently, the utilization of carbon-rich ash as biochar has become a topic of research and debate. It appears that those ashes where the carbon matrix has largely retained its original porous structure are most suitable. The high surface area may bind nutrients and the cavities may retain water or form a place where microorganisms can grow. This is one of the proposed mechanisms behind the success of Terra Preta [20].

10.4.3 Fuel

Utilization as fuel is the most obvious solution for carbon-rich ashes. Most of them still have a significant caloric value. It is important to keep in mind that the ash contains relatively large amounts of pollutants, so combustion facilities must have adequate flue gas cleaning. Also, fly ash is regarded as a waste material[11], and thus an installation that uses gasification ash as a fuel must comply with regulations for a waste incinerator. This does not apply, when the combustor is integrated with the gasifier, so that the facility can be regarded as a single installation.

Combustion will produce a certain amount of ash for which (again) a solution must be found. However, the fly ash is then nearly identical to fly ash from a combustion installation and can be regarded as such. In practice, the key questions are to find a buyer for the ash, establish a price, and find a solution for the combustion ash.

10.4.4 Other Applications for Carbon-rich Ashes

In the GASASH project, various alternative options have been investigated [21]. Some of them are briefly mentioned here.

Metallurgical applications are technically feasible, where the carbon-rich fly ash replaces part of the cokes or fossil carbon used to reduce metal ores. This kind of application requires large volumes of fly ash with a highly predictable quality. In reality, carbon-rich fly ashes with consistent quality are not (yet) on the market in substantial quantities.

Ashes that can bind considerable amounts of water can be used in fire retarding materials [22]. The use of carbon in fire retardant materials is counterintuitive, but

---

[11] Carbon-rich fly ash is likely to be categorized as non-hazardous waste, EURAL code 10 01 17. The gasification fly ash from demolition wood is also classified as non-hazardous, EURAL code 19 01 14.

in this application the carbon is used to bind water. It will only ignite once the water has evaporated[12].

Carbon-rich fly ashes originating from clean wood may be compressed into briquettes and used in barbecues and/or fire-places.

Carbon-rich ashes may be used to make synthetic basalt, fire-proof stones, or insulation materials. Preferably, the gasification ash should be combined with other (waste) materials to optimize the manufacture of synthetic basalt (viscosity, melting point, *etc.*). High temperature processes are expensive due to the energy consumption. However, energy consumption can be lowered[13] when the carbon in the ash is used as such in the manufacture of these products.

Just as in the case of fly ashes from biomass combustion, consistency and quality control are of key importance. Niche applications typically involve products with a high added value, where the use of good quality raw materials is of high importance. Carbon-rich fly ash is only attractive for replacing more expensive virgin materials if it has a guaranteed, predictable, and consistent quality.

References

1. KEMA (1998) Zekerstelling van kolenreststoffen. Fijnmazige selectie van toepassings-mogelijkheden. Report 98570110-KST/MAT 98-6608, KEMA, Arnhem, Netherlands (in Dutch)
2. Pels JR *et al.* (2004) Askwaliteit en toepassingsmogelijkheden bij verbranding van schone biomassa (BIOAS). Report ECN-C-04-091, ECN, Petten, Netherlands (in Dutch)
3. Pels JR *et al.* (2006) GASASH – improvement of the economics of biomass/waste gasification by higher carbon conversion and advanced ash management. Report ECN-C-06-038, ECN, Petten, Netherlands
4. Sarabèr AJ, Overhof LFAG (2009) Identification of bulk and niche applications of ashes derived from co-combustion, co-gasification and biomass combustion Report 50780586.610-TOS/ECC 08-9280 (confidential), KEMA, Arnhem, Netherlands
5. Gómez-Barea A *et al.* (2009) Plant optimisation and ash recycling in fluidised bed waste gasification. Chem Eng J 146:227–236
6. Communication from the Commission to the Council and to Parliament (1989) A community strategy for waste management. SEC (89):934 final, 18 September 1989
7. Sarabèr AJ *et al.* (2008) Prediction of volume and composition of future ashes in the Netherlands, Report, 50780586.610-TOS/MEC 08-9086 (confidential), KEMA, Arnhem
8. Dutch Building Materials Decree (Bouwstoffenbesluit) (1995) Nr. 12689/165, Netherlands Ministry of Housing, Spatial Planning and the Environment (VROM)
9. Schreurs JPGM, www.besluitbodemkwaliteit.info, last access date: 2010 08 30

---

[12] Even dry carbon-rich fly ash is difficult to ignite. It is completely free of volatiles and requires an igniter fuel, just like charcoal in barbecues.

[13] The costs figures of €200–350 per ton of ash, quoted in the report of the GASASH project [21], are based on existing processes for smelting oxidized ashes, notably the processes used in waste management in Japan. When the carbon in the fly ash substitutes a large part of the fuel needed to reach the melting temperature, much lower numbers are reached of €50–100 per ton of ash [23].

10. Proceedings of the RECASH International Seminar, 8–10 November 2005, Prague, Czech Republic, published by Forests of the Czech Republic, ISBN 80-86945-10-3; and proceedings of the RECASH 2nd International Seminar, 26–28 September 2006, Karlstad, Sweden, published by the Swedish Forestry Agency

11. Skogsstyrelsen (2001) Rekommendiationer vid uttag av skogsbränsle och kompensationsgödsling (Recommendations for the extraction of forest fuel and compensation fertilizing Swedish Forestry Agency (in Swedish)

12. Skogsstyrelsen (2001) Skogsvårdslagen-Handbok (Silvicultural Act – Handbook), Swedish Forstry Agency (in Swedish)

13. Emilsson S (2006) International handbook. From extraction of forest fuels to ash recycling. Swedish Forestry Agency ISBN 91-975555-1-7

14. Moolenaar SW, De Haas MJG (2004) Landbouwkundige beoordeling biomassa-assen ECN (Agricultural assessment of biomass ashes ECN), Report 948.04, NMI, Wageningen, Netherlands (in Dutch)

15. Nutrienten Management Instituut (2000) Handboek Meststoffen (Handbook Fertilizers), ISBN 90 5439 096 4, NMI, Netherlands

16. Sear LKA (ed) (2006) Proceedings of AshTech 2006, the international conference organized by the UKQAA, 15–17 May 2006, Birmhingham, UK, ISBN 0-9553490-0-1 (on CD)

17. Relini G (2000) Coal ash for artificial habitats in Italy. In: Jensen AC et al. (eds) Artificial reefs in European seas. Kluwer Academic Publishers, Dordrecht, The Netherlands

18. Kirbo K, www.reefball.org, last access date: 2010 08 15

19. www.c-fix.com, last access date: 2010 09 03

20. Day D (2005) Economical carbon capture, storage and utilization while producing renewable hydrogen and transportation fuels. Proceedings of the 14th European Biomass Conference and Exhibition, 17–21 October 2005, Paris, France, ISBN 88-89407-07-7 (on CD)

21. GASASH Deliverable Report 20 (2002) Improvement of the economics of biomass/waste gasification by higher carbon conversion and advanced ash management, ENK5-CT-2002-00635

22. Leiva C et al. (2007) Use of biomass gasification fly ash in light weight construction boards. Energ Fuel 21:361–367

23. Boersma AR et al. (2006) Smelten van biomassa-reststromen en afval tot (duurzame) energie en grondstoffen: een onderzoek naar mogelijke brandstofmengsels en een smeltervoorontwerp, Report ECN-C-06-013, ECN, Petten, Netherlands (in Dutch)